# Assessment of Chloramine and Chlorine Residual Decay in the Distribution System

Awwa Research Foundation

## About the Awwa Research Foundation

The Awwa Research Foundation (AwwaRF) is a member-supported, international, nonprofit organization that sponsors research to enable water utilities, public health agencies, and other professionals to provide safe and affordable drinking water to consumers.

The Foundation's mission is to advance the science of water to improve the quality of life. To achieve this mission, the Foundation sponsors studies on all aspects of drinking water, including supply and resources, treatment, monitoring and analysis, distribution, management, and health effects. Funding for research is provided primarily by subscription payments from approximately 1,000 utilities, consulting firms, and manufacturers in North America and abroad. Additional funding comes from collaborative partnerships with other national and international organizations, allowing for resources to be leveraged, expertise to be shared, and broad-based knowledge to be developed and disseminated. Government funding serves as a third source of research dollars.

From its headquarters in Denver, Colorado, the Foundation's staff directs and supports the efforts of more than 800 volunteers who serve on the board of trustees and various committees. These volunteers represent many facets of the water industry, and contribute their expertise to select and monitor research studies that benefit the entire drinking water community.

The results of research are disseminated through a number of channels, including reports, the Web site, conferences, and periodicals.

For subscribers, the Foundation serves as a cooperative program in which water suppliers unite to pool their resources. By applying Foundation research findings, these water suppliers can save substantial costs and stay on the leading edge of drinking water science and technology. Since its inception, AwwaRF has supplied the water community with more than $300 million in applied research.

More information about the Foundation and how to become a subscriber is available on the Web at **www.awwarf.org**.

# Assessment of Chloramine and Chlorine Residual Decay in the Distribution System

Prepared by:
**Zaid K. Chowdhury** and **Laurel Passantino**
Malcolm Pirnie, Inc., Phoenix, AZ 85008

**R. Scott Summers**, **Lori Work**, and **Natalie Smith**
University of Colorado, Boulder, CO 80309

**Lewis Rossman**
United States Environmental Protection Agency, Cincinnati, OH 45268

and

**James Uber**
University of Cincinnati, Cincinnati, OH 45221

Sponsored by:
**Awwa Research Foundation**
6666 West Quincy Avenue, Denver, CO 80235-3098

Published by:

Awwa Research Foundation

American Water Works Association

IWA Publishing

## DISCLAIMER

This study was funded by the Awwa Research Foundation (AwwaRF). AwwaRF assumes no responsibility for the content of the research study reported in this publication or for the opinions of statements of fact expressed in the report. The mention of trade names for commercial products does not represent or imply the approval or endorsement of AwwaRF. This report is presented solely for informational purposes.

# CONTENTS

# TABLES

# FIGURES

# FOREWORD

The Awwa Research Foundation (AwwaRF) is a nonprofit corporation that is dedicated to the implementation of a research effort to help utilities respond to regulatory requirements and traditional high-priority concerns of the industry. The research agenda is developed through a process of consultation with subscribers and drinking water professionals. Under the umbrella of a Strategic Research Plan, the Research Advisory Council prioritizes the suggested projects based upon current and future needs, applicability, and past work; the recommendations are forwarded to the Board of Trustees for final selection. The foundation also sponsors research projects through an unsolicited proposal process; the Collaborative Research, Research Applications, and Tailored Collaboration programs; and various joint research efforts with organizations such as the U.S. Environmental Protection Agency, the U.S. Bureau of Reclamation, and the Association of California Water Agencies.

This publication is a result of one of these sponsored studies, and it is hoped that its findings will be applied in communities throughout the world. The following report serves not only as a means of communicating the results of the water industry's centralized research program but also as a tool to enlist the further support of the nonmember utilities and individuals.

Projects are managed closely from their inception to the final report by the foundation's staff and large cadre of volunteers who willingly contribute their time and expertise. The foundation serves a planning and management function and awards contracts to other institutions such as water utilities, universities, and engineering firms. The funding for this research effort comes primarily from the Subscription Program, through which water utilities subscribe to the research program and make an annual payment proportionate to the volume of water they deliver and consultants and manufacturers subscribe based on their annual billings. The program offers a cost-effective and fair method for funding research in the public interest.

A broad spectrum of water supply issues is addressed by the foundation's research agenda: resources, treatment and operations, distribution and storage, water quality and analysis, toxicology, economics, and management. The ultimate purpose of the coordinated effort is to assist water suppliers to provide the highest possible quality of water economically and reliably. The true benefits are realized when the results are implemented at the utility level. The foundation's trustees are pleased to offer this publication as a contribution toward that end.

Walter J. Bishop
Chair, Board of Trustees
Awwa Research Foundation

Robert C. Renner, P.E.
Executive Director
Awwa Research Foundation

# ACKNOWLEDGMENTS

The authors of this report are indebted to the following water utilities and individuals for their cooperation and participation in this project:

- City of Boulder, Colorado
- City of Newport News, Virginia
- City of Scottsdale, Arizona
- City of Winchester, Kentucky
- Greater Cincinnati Water Works
- Hillsborough County Water District

In addition, the authors would like to thank two individuals that were essential during the early stages of the project: Albert Ilges of AwwaRF, who served as the first project officer, and Mr. Warren Swanson of Schmueser Gordon Meyer, Inc. (formerly of Malcolm Pirnie, Inc.), who coordinated the utility participation.

Lastly, the authors would like to thank the AwwaRF project officer, Ryan Ulrich, and all the members of the Project Advisory Committee: Kanwal Oberoi, Timothy Perkins, Janice Skadsen, and John Vasconcelos.

# ACKNOWLEDGEMENTS

# EXECUTIVE SUMMARY

## BACKGROUND

Maintaining chlorine and chloramine residual is a key factor that determines potability of distributed water. Decay of these disinfectants is affected by various water quality and infrastructure related factors. Understanding the decay of residuals in the distribution system is critical with respect to safe delivery of drinking water and in developing water quality models that can be used by utilities for the operation of potable water delivery systems. In order to comply with microbial and disinfection byproduct (M/DBP) regulations, water suppliers are continually investing in treatment processes for achieving greater levels of inactivation and DBP precursor removal. As a result of these changes in water quality, maintaining a disinfectant residual in the distribution system can be challenging.

The primary objective of this research was to determine the effect of advanced treatment on chlorine and chloramine residual decay in the distribution system, considering the bulk phase disinfectant demand as well as the pipe-wall demand. Additional objectives included determining the effect of dose, temperature, pH, blending, and rechlorination on disinfectant decay, developing models to predict disinfectant decay, and updating the algorithms in the EPANET Programmer's Toolkit to provide more robust water quality and hydraulic modeling capability.

## APPROACH

The research approach followed in this study addressed the issues of chlorine and chloramine decay in bench-, pilot-, and field-scale applications following conventional and advanced treatment. In the bench-scale testing, the effect of advanced treatment (enhanced coagulation [EC], granular activated carbon [GAC] adsorption, ozonation [$O_3$], biofiltration [BF], and ultraviolet [UV] irradiation) on chlorine and chloramine decay was evaluated for three water sources (City of Boulder, Colorado; City of Newport News, Virginia; and City of Winchester, Kentucky). Additional studies examined the effects of disinfectant dose, temperature, pH, blending, and rechlorination.

The pilot-scale testing used a pipe loop distribution system simulator at the United States Environmental Protection Agency's Test and Evaluation Center in Cincinnati, Ohio. The water used in these evaluations was partially treated Ohio River water that had been coagulated, settled, and filtered (but not chlorinated) at the Greater Cincinnati Water Works. The advanced treatments applied in these studies were GAC, $O_3$, and reverse osmosis (RO). The effect of these treatments on chlorine and chloramine decay was evaluated at two disinfectant doses.

The field-scale testing was performed in three cities: City of Newport News, Virginia; City of Scottsdale, Arizona; and "Utility A," located in the southeastern United States. Field testing in Newport News involved monitoring chloramine residual at two locations within the distribution system. The locations were selected so that each one received water from only one of the City's water treatment plants (WTPs). In Scottsdale, a small portion of the distribution system was selected to perform a tracer study in conjunction with chlorine residual monitoring. The results from the tracer study provided travel time. When coupled with the chlorine residual data, this provides an assessment of decay as a function of time in this area. The most complete data set and detailed analyses were gathered for Utility A. At this field test site, a system-wide tracer study was performed using two independent tracers (conductivity and fluoride). The data from this study were used to calibrate the hydraulic model which in turn allowed for calibration and verification of the model extensions developed in this project: network path analysis and multi-species chlorine decay.

## CONCLUSIONS

The evaluations performed at the bench- and pilot-scale indicated that advanced treatment affects residual decay. For the bench-scale studies, GAC and $O_3$/BF yielded the smallest disinfectant demand for the bulk water (or most stable disinfectant residual). However, on the pilot-scale, the ozonated and GAC-treated waters resulted in the highest decay rates for the pipe-wall coefficient. Both components need to be considered when determining overall disinfectant decay in the distribution system. The effects of disinfectant dose, temperature, pH, blending, and rechlorination on residual decay were also examined, and the following results were noted:

- Disinfectant Dose: An increase in chlorine dose resulted in an increased chlorine demand. However, an increase in chloramine dose did not affect chloramine demand.
- Temperature: Temperature effects were most prevalent in the latter part of the decay curves (after 3 and 8 hours for chlorine and chloramine decay, respectively).
- pH: Disinfectant decay was affected by changes in pH. However, the results were not systematic.
- Blending: Blending water with different chlorine residuals resulted in a more stable chlorine residual. Blending of chloraminated waters resulted in a chloramine residual that was equal to the weighed average.
- Rechlorination: The residual decay after rechlorination depends only on the total disinfectant dose applied to the water without consideration of how the total dose is divided and applied.

The bench-scale data were used to develop simple models for predicting chlorine and chloramine decay as a function of total organic carbon (TOC), ultraviolet absorbance at 254 nm (UVA), and disinfectant dose.

Lastly, the updated algorithms in EPANET are a valuable addition to the water quality modeling software currently available. The updated algorithms account for chemical reactions between chlorine and other species in the distribution system (such as natural organic matter or ammonia) and also account for the multiple paths water can take as it travels from an input to an output location.

# CHAPTER 1
# INTRODUCTION

## PROBLEM DEFINITION

To provide adequate protection of public health, surface waters need to be disinfected (primary disinfection) at the water treatment plant (WTP) and distributed to customers while maintaining a disinfecting environment (residual disinfection). In order to achieve compliance with stricter microbial and disinfection byproduct (M/DBP) regulations, water suppliers are continually investing in treatment processes for achieving greater levels of inactivation and DBP precursor removal at the WTP. The requirements of the Total Coliform Rule (TCR) also mandate water utilities monitor chlorine residual throughout the distribution system and ensure that a measurable residual persists in all areas of the system. The next generation of this rule (under development at the time of writing this report) is expected to be more encompassing and is being tentatively titled Distribution System Rule. As a result of these changing dynamics of water quality and regulatory environment, the maintenance of residual in the distribution system often poses a challenge to the water systems. Understanding the decay of residuals in the actual distribution system is critical with respect to developing water quality models that can be used by utilities for the operation of potable water delivery systems.

In the United States, chlorine is the most commonly used disinfectant for both primary and residual disinfection. However, the use of chloramines for residual disinfection is increasing. A significant increase in chloraminated systems is anticipated as water utilities begin to implement changes to comply with the Stage 2 D/DBP Rule. The decay of chlorine and chloramine residuals in the water distribution system depends on many factors. Although, it is generally understood that disinfectant demand is exerted by bulk water quality parameters and pipe materials, no systematic evaluation of these influences has been performed. Current models that attempt to describe residual decay in the distribution system are generally poor predictors of chlorine in the distribution system. The decay kinetics of residuals following advanced treatment at the WTP are even lesser known compared to decay kinetics followed by conventional treatment.

The research approach followed in this study addressed the issues of chlorine and chloramine decay in bench-, pilot-, and field-scale applications following conventional and advanced treatment. In these experiments, comparing laboratory experiments with the pipe-loop (pilot-scale) and field studies isolated the effects of bulk and pipe-wall decay. The data collected during the study were used to develop more comprehensive models of residual decay compared to existing models.

## RESEARCH OBJECTIVES

The objectives of this research were as follows:

- determine the effect of advanced treatment on chlorine and chloramine residual decay in the distribution system, considering the bulk phase disinfectant demand as well as the pipe-wall demand

- determine the effects of chlorine dose, temperature, pH, blending, and rechlorination on disinfectant decay
- develop models to predict disinfectant decay as a function of basic water quality parameters
- develop more robust algorithms for disinfectant decay in the distribution system to include in the EPANET programmer's toolkit

## RESEARCH BENEFITS

The benefit of this research is that the results provide a framework for utilities to assess and predict disinfectant decay in the distribution system. The chlorine and chloramine decay models developed as a result of the bench-scale testing provide a simple method of predicting changes in disinfectant decay that result from changes in treatment strategy. However, bulk demand alone is not sufficient to characterize decay in the distribution system. The demand at the pipe-wall must also be considered. This research also provides information on how changes in treatment strategy affect the pipe-wall demand coefficients. Lastly, the hydraulic and water quality modeling aspects of this research allow for detailed chlorine decay information at all points in the distribution system. The updated algorithms developed for this project account for multi-species water quality modeling and also account for the multiple paths water can take from an input location to an output location.

# CHAPTER 2
# LITERATURE REVIEW

## CHLORINE AND CHLORAMINE DECAY KINETICS IN BULK WATERS

Chlorine and chloramine decay kinetics have been found to be a function of both water quality parameters and operational conditions. Under typical water quality conditions, natural organic matter (NOM) serves as the main reactant for these disinfectants. NOM can be measured by many different parameters, such as total organic carbon (TOC) concentration, dissolved organic carbon (DOC), or ultraviolet absorbance (UVA) measured at wavelength of 254 nanometers. The specific UVA:DOC ratio (SUVA) is often used to describe the aromatic character of NOM. Other water quality parameters that have been shown to affect chlorine and chloramine decay are inorganic compound concentrations, temperature, and pH. Operational conditions such as disinfectant dose and water treatment processes also can affect chlorine and chloramine decay kinetics. This section is a review of available information on the effect of these parameters and conditions on chlorine and chloramine decay, as well as a summary of existing chlorine and chloramine decay models.

### Chlorine Chemistry

#### *Chemical Reactions With Organic Compounds*

Chlorine reacts with organic compounds via three pathways: 1) oxidation of organic functional groups, 2) addition to double bonds and 3) electrophilic substitution (Brezonik 1994). The majority of chlorine is consumed via the oxidation pathway, while the latter two produce halogenated DBPs. All compounds containing chlorine in oxidation states greater than -1 are unstable with respect to redox equilibria in aqueous solution. These compounds act as oxidizing agents for organic matter (Brezonik 1994). In oxidation reactions, chlorine species are reduced to chloride ions. Chlorinated organic compounds are not formed in oxidation reactions with organic matter. Unfortunately, complete oxidation to $CO_2$ is uncommon, and intermediate organic products such as aldehydes and carboxylic acids are formed (van Hoof 1992). However, at a relatively high chlorine dose of approximately 10 mg/L, Helz et al. (1983) reported that $CO_2$ yield accounted for about half of the total observed chlorine decay in a river water. The time scale of chlorine oxidation reactions varies widely depending on the nature of the organic compound. Reactions with organic sulfur and some aromatic and nitrogen heterocyclic rings are very fast ($t_{1/2}$ < 1 sec). Oxidation of amines occurs on a time scale of minutes to hours, and nonselective oxidation of organic matter occurs over minutes to days (Dotson and Helz 1985).

#### *Chemical Reactions With Inorganic Compounds*

Chlorine is a strong oxidant and can react with inorganic compounds found in source waters (e.g., ammonia, iron and manganese). Free chlorine reacts rapidly with ammonia to form different species of chloramines. At $Cl_2/N$ weight ratios greater than 7.6 (breakpoint at neutral pH), excess free chlorine is available, which can then react with organic matter to form DBPs. Chlorine will oxidize soluble iron ($Fe^{2+}$) and manganese ($Mn^{2+}$) to the insoluble ferric and manganic forms, respectively. The theoretical stoichiometric requirements for these reactions are 0.64 mg free chlorine per mg iron and 1.3 mg free chlorine per mg manganese. The iron oxidation will take place over a wide range of pH (4-10) with an optimum at pH 7. The reaction is usually complete after one hour. Manganese reacts slowly with chlorine at pH values below 8 (White 1992). Chlorine also oxidizes hydrogen sulfide. The chemistry is extremely complex with theoretical stoichiometric requirements of 2.1 to 8.5 mg chlorine for each mg $H_2S$ (White 1992). While chlorine reacts with bromide and the reaction is important in the formation of brominated DBPs, the bromide concentrations are typically low (less than 0.1 mg/L) thus their contribution to chlorine demand is low.

### Chlorine Decay Kinetics

Chlorine decay kinetics depend on chlorine dose, treatment processes (which affect NOM concentration and character), pH, and temperature. Longer reaction times lead to higher chlorine consumption and greater DBP formation. The majority of the chlorine, however, is consumed in the early part of the reaction. Summers et al. (1996) showed that approximately 80 to 85 percent of the 3-day chlorine demand was consumed after 24 hours in medium TOC water (2.2 to 4.1 mg/L). In both cases, the free chlorine residual after 24 hours was targeted at 1 mg/L.

#### *Effect of Chlorine Dose*

Several studies have shown that chlorine demand increases with increasing $Cl_2$/DOC dose ratios (Shukairy et al. 1994; Fleischacker and Randtke 1983). However, the demand per mg chlorine dose decreases. This suggests that the reaction becomes precursor limited with increasing dose. Vasconcelos et al. (1995) described chlorine decay in distribution systems with a first-order reaction model and reported that lower initial chlorine concentrations lead to higher rates of chlorine consumption.

Most of chlorine decay work thus far has been with a constant chlorine dose to TOC ratio, expressed as $Cl_2$:TOC in units of mg/mg. However, the initial concentration of chlorine has been shown to affect the residual concentration (Dugan et al. 1995; Matheis et al. 1998; Koechling 1998; Isabel et al. 2000). Koechling (1998) showed that the chlorine dose did not affect the chlorine decay kinetics when expressed as a percent of the overall demand for the raw waters tested. In other words, the relative chlorine consumption kinetics were the same at either dose for the entire reaction period. None of the above studies had an adequate database for specifically evaluating the effect of dose on short-term chlorine decay.

The Koechling (1998) model was developed from waters dosed at a 2:1 $Cl_2$:TOC ratio. Isabel et al. (2000) attempted to amend the rate constants to account for other dose ratios, but they only used five waters and the result was highly empirical. The two species model (McClellan et al. 2000; Clark and Sivaganesan 2002; and Boccelli et al. 2003) includes chlorine dose as an initial parameter and assumes that the rate constants are the same, regardless of dose. However, more waters need to be tested to determine if the model holds true at a large dose range.

#### *Effect of Treatment and NOM*

Most chlorine decay studies have been conducted on raw waters or conventionally treated waters. Coagulation is a widely used process in drinking water treatment. It can lower the TOC concentration and UVA significantly, which leads to a removal of chlorine demand. Generally, coagulants are more effective in removing higher molecular weight hydrophobic matter than lower molecular weight hydrophilic matter (Edzwald 1994). A few researchers have investigated the effect of coagulation, flocculation, and sedimentation on long-term chlorine decay (Matheis et al. 1998; Koechling 1998; and Isabel et al. 2000). The process decreases TOC concentration and SUVA, and subsequently the chlorine decay rate. When the waters were dosed at the same $Cl_2$:TOC dose ratio, the normalized chlorine decay rate (chlorine residual divided by chlorine dose) tended also to decrease with coagulation in most waters, but not always. This indicates that coagulation does not preferentially remove chlorine reactive fractions of the NOM.

Ozonation oxidizes NOM, thereby breaking down larger molecules into smaller ones, and increasing the polarity and hydrophilicity of the NOM. Ozone reacts with unsaturated and aromatic functional groups in NOM to form nonhalogenated by-products such as aldehydes, ketones and carboxylic acids (Langlais et al. 1991). Ozonation was shown to decrease UVA, thus it destroys some of the reactive sites which then are no longer available for subsequent chlorination reactions. Therefore, chlorine demand decreases after ozonation (Miltner et al. 1992; Shukairy et al. 1995). Koechling (1998) and Matheis et al. (1998) have found that pre-ozonation sometimes reduces the long-term chlorine decay. However, more waters need to be studied to confirm the findings. Ozonation followed by biofiltration has been shown to greatly decrease the chlorine decay rate relative to the TOC (Matheis et al. 1998; Koechling 1998; and Isabel et al. 2000). This indicates that the ozone/biofiltration process preferentially removes chlorine reactive NOM.

To date, there is very little information on the effect of GAC adsorption on chlorine decay. Matheis et al. (1998) studied the effect on two waters and found conflicting results. In one case, the long-term chlorine demand of the TOC increased after GAC treatment, but in the other case it decreased.

Ultraviolet (UV) irradiation is an alternative to chemical disinfection, but after UV irradiation, a chemical disinfectant must be added to maintain a residual in the distribution system. The effect of UV irradiation on subsequent chlorine decay has never been investigated, though it has been shown that the process has no effect on subsequent DBP formation (Malley et al. 1995). In one study, nanofiltration was found to decrease the SUVA at 254 nm for some of the five waters, but increased it for others (Isabel et al. 2000). Therefore it did not consistently affect the normalized chlorine residual decay rate. Nanofiltration did, however, lower the DOC of the water and consequently lowered the decay rate.

The Koechling (1998) model was developed based on raw waters. It was later amended by Isabel et al. (2000) to account for changes in the rate constants due to treatment, but the results were highly empirical. The two-species model was tested on waters treated by a wide range of treatments (Clark and Sivaganesan 1998). However no one has specifically evaluated its performance at different treatment levels.

### *Effect of Temperature*

The temperature of some surface water supplies may vary over a range of 30°C or more during the course of a year (Fleischacker and Randtke 1983), and such temperature fluctuations are known to have a significant effect on disinfectant reactions. A study of chlorine decay in distribution systems reported, that for a temperature range of 5 to 25°C, the first-order chlorine decay constant varied between $1\times10^{-3}$ and $9\times10^{-3}$ $min^{-1}$ (Vasconcelos et al. 1995). An increase in temperature caused an increase in chlorine demand, particularly in the early stage of the reaction (Van Hoof 1992; Shukairy 1994; Summers et al. 1996). Koechling (1998) evaluated the effect of temperature on the chlorine decay of four raw waters and found, as expected, that lower temperatures suppressed the chlorine decay. A Van Hoof-Arrhenius relationship between rate constants at different temperatures was shown to be applicable to the Koechling (1998) model.

### *Effect of pH*

A poorly understood factor in chlorine decay is pH. For conventional drinking water treatment a chlorine/chloramine disinfection approach typically involves the application of chlorine prior to the rapid mix where the pH is 7.5 to 8 or after coagulation or sedimentation where the pH can drop below 6. Then, the pH is typically raised above 8 before release to the distribution system to minimize corrosion. If softening treatment is used, the pH can increase above 11.5 after lime addition. Then it must then be lowered to below 10 prior to the distribution system. The effects of different pHs on chlorine reaction kinetics have not been extensively evaluated. For chlorine reactions, a decrease in pH from 7 to 4.5 did not significantly change chlorine demand (Shukairy 1994) or for the pH range of 10 to 5 (Fleischacker and Randtke 1983). Koechling (1998) evaluated the effect in three waters, but the effect was not constant from water to water.

## Chlorine Decay Models

Many chlorine decay models have been developed over the past decade. A summary of some of them with their constraints is shown in Table 2.1.

Many researchers have modeled chlorine decay with first-order reactions (Vasconcelos et al. 1996). These models assume that the reactive component in the water is in large excess, and the reaction only depends on chlorine concentration. The rate coefficient, $k$, is depends highly on the source water and degree of treatment received. The results have been satisfactory when applied to distribution systems and the first few hours of decay data are ignored. Vasconcelos et al. (1995) have reviewed four kinetic models to describe the chlorine decay in distribution systems: (1) a first-order expression, (2) an-nth order equation, (3) limited first-order kinetics, and (4) a parallel first-order decay rate. Haas and Karra (1984) used a parallel first-order decay model to predict chlorine decay in wastewaters. This model assumes that two groups of water constituents react with chlorine: (1) fast reacting components, which exert some initial chlorine decay, and (2) slow reacting components, which are responsible for long-term chlorine demand.

USEPA's WTP Model Version 1.5 used a three component chlorine decay model. It describes the first five minutes of decay with a zero-order component and the short-term decay with a second-order model. After five hours, the long-term decay is described by a first-order reaction (MWDSC et al. 1991). The three rate consatants were found to be related to the initial $Cl_2$/TOC and UVA of the water. The model is not continuous and does not include any parameters representing the reactive species in the water.

Jadas-Hecart et al. (1992) used an initial phase of immediate chlorine consumption during the first 4 hours, and a second slow consumption phase after 4 hours for the long-term chlorine demand. However, this model cannot be applied from time zero of chlorination because the long-term demand obeys a different reaction order than the initial demand (before 4 hours). No attempt was made to predict the kinetic constant, and no model was suggested for the initial chlorine demand.

**Table 2.1**
**Recent chlorine decay model development**

| Study | Model | Parameters | Constraints |
|---|---|---|---|
| Jadas-Hecart et al. (1992) | $-\frac{dx}{dt} = k(a-x)^{\alpha}\left(b-\frac{x}{n}\right)^{\beta}$ | x = 4-hour chlorine demand<br>a = 4-hour chlorine residual<br>b (unknown value) = the content in precursors of long-term chlorine demand<br>k = rate constant<br>n = stoichiometry<br>$\alpha$, $\beta$ = partial orders of reaction | Long-term chlorine demand only. Experimentally determined short-term demand, residual and rate constant. |
| Michaelis-Menten (Koechling 1998) | $\frac{dC}{dt} = \frac{k_2 \cdot SUVA_0 \cdot C}{\alpha_1 + C}$<br>Integrated form:<br>$C_t = \left[a_1 \cdot \ln\left(\frac{C_0}{C_t}\right)\right]$<br>$-(k_2 \cdot SUVA_{254,0} \cdot t) + C_0$ | $C_t = Cl_2$ concentration at time, t<br>$C_0$ = initial $Cl_2$ concentration<br>$\alpha_1 = 4.98 \cdot UV_{254,0} - 1.91 \cdot TOC_0$<br>$r^2 = 0.999$, n = 11<br>$k_2 = -0.32 \cdot UV_{254,0}{}^{1.37}$<br>$r^2 = 0.882$, n = 11 | Not accurate at short-term time periods.<br>Based on initial water quality parameters, does not track decay of reacting species.<br>Based on single set of reaction conditions: (untreated, 2:1 $Cl_2$:TOC dose ratio, pH 8, 20°C (temperature correction developed) |
| Isabel et al. (2000) | (Same as Koechling 1998) | For 2:1 $Cl_2$:TOC dose ratio:<br>$\alpha_{1@2} = 4.98 \cdot UV_{254,0} - 1.91 \cdot TOC_0$<br>For $Cl_2$:TOC dose ratios ranging from 1:1 to 3:1):<br>$\alpha_1 = \alpha_{1@2} \cdot 0.503 \cdot \left(\frac{C_0}{TOC_0}\right)$<br>For all dose ratios and treatments:<br>$k_2 = -a \cdot UV_{254,0}^{b}$<br>where a and b depend on dose and treatment | Not accurate at short-term time periods.<br>Based on initial water quality parameters, does not track decay of reacting species.<br>Dose range 1:1 to 3:1 $Cl_2$:TOC.<br>Based on single pH and temperature (pH 8, 20°C).<br>Limited data set. |
| Two-Species Model (Clark 1998) | $\frac{dC_A}{dt} = k_A C_A C_B$<br>$\frac{dC_B}{dt} = k_B C_A C_B$ | $C_A$, $C_B$ = free chlorine and reacting NOM concentrations<br>$k_A$, $k_B$ = second-order rate constants | Reactants are fictive and not measurable.<br>Fictive reactive concentration and rate constants must be determined experimentally. |
| McClellan et al. (2000) | $\frac{dCl_2}{dt} = n_1 k_1 [S_1][Cl_2]$<br>$+ n_2 k_2 [S_2^-][HOCl]$<br>[ ] denotes concentration | $S_1$, $S_2^-$ = fictive reactive species<br>* $S_2^-$ (deprotonated) concentration is pH dependent<br>$n_1$, $n_2$ = coefficients that account for $Cl_2$ consumption in intermediate steps<br>$k_1$, $k_2$ = second-order rate constants | Reactants are fictive and not measurable.<br>Fictive reactive concentration and rate constants must be determined experimentally. |

Dugan et al. (1995) and Koechling (1998) developed a Michaelis–Menten saturation type model to describe chlorine decay. The Koechling (1998) model attempts to predict the entire chlorine decay reaction from initial water quality parameters (TOC, UVA, and SUVA). The model is simple, and it accurately predicts long-term decay. Although the model is based on initial water quality conditions, it does not include a true measure of the reacting components in the water and their decrease with chlorine reaction time. The model fails to predict short-term chlorine decay.

Isabel et al. (2000) performed a study of five water sources to evaluate the effect of chlorine dose and treatment (coagulation, ozonation/biofiltration, and nanofiltration) on chlorine decay. They extended the Koechling (1998) model to represent the decay of treated waters and account for a practical range of chlorine doses. They found that as the $Cl_2$:TOC ratio increased, so did the chlorine demand of the TOC. This effect became less pronounced at very high dose ratios, indicating, like the results of Matheis et al. (1998), that at high $Cl_2$:TOC ratios the reaction is NOM limited, whereas at lower dose ratios the reaction is chlorine limited. They found the model of Koechling, (1998) worked for treated waters and waters dosed at a practical range of $Cl_2$:TOC ratios (1:1 to 3:1). They developed a multiplier for $k_1$ (Table 2.1) in the cases of dose ratios other than 2:1 $Cl_2$:TOC and found that $k_2$ could be related to UVA for each level of treatment by a power function (Table 2.1). The USEPA WTP Model Version 2.0 uses the approach of Koechling (1998) as modified by Isabel et al. (2000).

Clark (1998) investigated a two-species, second-order overall model, in which chlorine decay reactions are first-order with respect to chlorine and first-order with respect to a fictive reactive component. This model is similar to an earlier model by Jadas-Hecart et al. (1992). The model was found to be more successful than the first-order model at describing long-term chlorine decay. Clark and Sivaganesan (1998) developed empirical power function equations to predict the model parameters based on initial chlorine concentration, TOC, temperature, and pH. The two-species model simulates a reacting fraction of NOM, so it can predict chlorine decay after rechlorination significantly better than the first-order decay model (Boccelli et al. 2003).

Clark and Sivaganesan (2002) developed a model based on two separate two-species reactions in parallel. They used the data set of Koechling (1998) to develop the model. One reaction represents the chlorine decay of fast-reacting species, and the other describes slow-reacting species. Empirical power function equations to predict the model parameters are based on initial chlorine concentration, UVA, TOC, temperature, pH, bromide concentration, and alkalinity.

McClellan et al. (2000) developed a mechanistic model to predict chlorine decay. It assumes a constant for the instantaneous chlorine demand and two hypothetical reaction sites, $S_1$ and $S_2$. The rate-limiting step for $S_1$ is a two-species second-order reaction. The rate-limiting step for $S_2$ is a proton dissociation reaction which is first-order with respect to the $S_2$ concentration. This reaction is followed by a faster two-species reaction with chlorine. The $S_2$ reaction was developed to mimic a known reaction pathway between chlorine and ketones. The McClellan et al. (2000) model and the two-reaction/two-species model (Clark and Sivaganesan 2002) appear to fit the entire chlorine decay curve. However, only a few data points were taken in the short-term time period. While these models and the two-species model include parameters that represent the concentration of reactive NOM species, no measurable surrogates have been found that actually reflect the decrease in these species.

## Chloramine Chemistry

Advantages of using chloramines as secondary disinfectants to maintain a distribution system disinfectant residual include the increased stability of residuals in the distribution system and a lower formation of regulated DBPs. Chloramines react relatively slowly with organic precursors (Wolfe et al. 1984). Thus, the overall extent of chlorinated by-product formation, especially trihalomethanes (THMs), is less than for free chlorine (Van Hoof 1992). Additionally, its slow reaction rate allows chloramine to penetrate into biofilms growing in distribution systems, thus increasing the inactivation of biofilm microorganisms. Disadvantages involved with the use of chloramines are the lower effectiveness for inactivation of bacteria, viruses, and protozoans. To meet primary disinfection requirements with chloramine, high CT values are needed, requiring either high concentrations or long detention times. For this reason, few utilities use it primary disinfection.

Chloramines are formed when ammonia and free chlorine react. The formation of the chloramines can be depicted as a stepwise process:

$$NH_{3(aq)} + HOCl \leftrightarrow H_2O + NH_2Cl \quad \text{(monochloramine)}$$

$$NH_2Cl + HOCl \leftrightarrow H_2O + NHCl_2 \quad \text{(dichloramine)}$$

$$NH_2Cl + HOCl \leftrightarrow H_2O + NCl_3 \quad \text{(trichloramine)}$$

The distribution among mono-, di-, and trichloramine depends on pH, temperature, chlorine to ammonia ratio ($Cl_2/N$), and the presence of other compounds, with the $Cl_2/N$ ratio being the most significant factor (Van Hoof 1992). Monochloramine predominates at pH > 7.5 (Van Hoof 1992) and at a $Cl_2/N$ weight ratio of 5 or smaller (Wolfe et al. 1984). It is the dominant species under practical operating conditions of distribution systems (high pH and high chlorine to nitrogen ($Cl_2$:N) ratios). Dichloramine formation is favored by pH decrease and a $Cl_2/N$ increase. Its maximum formation rate is between pH 4.4 and 6 and/or a $Cl_2/N$ weight ratio between 5 and 7.6. Trichloramine is formed at pH < 4.4 or at $Cl_2/N > 7.6$ (Wolfe et al. 1984). In drinking water treatment, reaction conditions are typically controlled through the $Cl_2/N$ ratio and pH control to ensure that monochloramine is the dominant species present. The formation of di- and trichloramine is undesirable because these compounds are thought to impart a disagreeable chlorinous taste at concentrations above 0.8 mg/L and 0.02 mg/L, respectively (Wolfe et al. 1984).

Monochloramine undergo hydrolysis according to the following equation to form HOCl (Jafvert and Valentine 1992):

$$NH_2Cl + H_2O \rightarrow HOCl + NH_3$$

Ammonia is important in chloramine reactions because of its effect on the $Cl_2/N$ ratio. At low ratios (<5 on a weight basis) and neutral to high pH, monochloramine is the dominant chloramine species. As the $Cl_2/N$ ratio increases, more dichloramine is formed, especially at low pH (Diehl et al. 1995).

## Chloramine Decay Kinetics

### *Factors Affecting Chloramine Decay Kinetics in the Absence of NOM*

Chloramines are unstable at neutral pH values as a result of "auto-decomposition." Auto-decomposition is a collective term that describes chemical reactions that occur among waters containing free chlorine and ammonia, which ultimately result in the oxidation of ammonia and the reduction of active chlorine. Valentine and co-workers (Jafvert and Valentine 1992; Ozekin et al. 1996; Vikesland et al. 2001) developed a monochloramine decay model in order to describe the process of auto-decomposition. Chloramine loss by auto-decomposition is a complex process. The model includes fourteen reactions, but at neutral pH and above, monochloramine decay is limited by dichloramine formation.

This monochloramine auto-decomposition model readily predicts monochloramine loss over an initial concentration range of 0.7 – 7.5 mg/L as $Cl_2$ (Vikesland et al. 2001). Monochloramine auto-decomposition was found to be neither a function of ionic strength nor initial concentration. However, pH shows a significant effect on monochloramine decay kinetics. An increase in monochloramine decay kinetics followed a decrease in pH values from 8.3 to 6.6. This accelerated loss of monochloramine can be attributed to the fact that the formation rate for dichloramine, which decomposes rapidly, is faster at lower pH values (Vikesland et al. 2001). The products of dichloramine decomposition are typically ammonia, chloride, and nitrogen gas. In addition, nitrate may form under certain conditions (Vikesland et al. 1998). As expected, the effects of monochloramine decay depend highly on temperature. Vikesland et al. (1998) conducted temperature studies that ranged from 4°C to 35°C. It was found that monochloramine is more stable at lower temperatures. Furthermore, the decay studies concluded that the half-life of monochloramine exceeds 300 hr at 4°C, while the half-life for monochloramine at 35°C was only 75 hr (Vikesland et al. 2001).

Monochloramine decay kinetics were also found to be a function of the $Cl_2$:N molar ratio. It is important for individual distribution systems to maintain an appropriate $Cl_2$:N ratio on a case by case basis. As $Cl_2$:N ratios increase, chloramine decay kinetics showed a slight increase at a pH of 7.55. This increase in chloramine decay kinetics is more significant at a lower pH value of 6.55 (Vikesland et al. 2001).

The effects of nitrite and bromide were investigated both individually and collectively. Experiments conducted to examine the effect of nitrite on monochloramine decay showed that levels of nitrite that might be found in certain drinking water distribution systems exert a significant demand on monochloramine. The effects of low levels of bromide (levels typically found in distribution systems) on monochloramine decay are insignificant. Increasing levels of bromide correspond to an increasing demand on monochloramine. As for the influence of both nitrite and bromide, it was found that these two constituents react synergistically and pose a greater demand on monochloramine collectively than was observed individually (Vikesland et al. 2001).

Little has been recently published on iron or manganese oxidation with chloramines. According to White (1992), chloramines will oxidize $Fe^{2+}$ at a stoichiometric requirement of 0.64 mg chloramine per mg iron, which is the same as for free chlorine. However, if the iron is in a complex organic form, free chlorine is more effective than combined in breaking up the iron complex so that oxidation by chlorine can proceed. For manganese, Shorney and Knocke (1996) showed that chloramines oxidized $Mn^{2+}$ at much lower rates than free chlorine.

When chloramine decay studies were conducted on natural waters from various sources, the chloramine decayed more rapidly than the model predicted. This implies that there is an extra demand imposed on chloramine due to NOM in real waters.

### *Factors Affecting Chloramine Decay Kinetics in the Presence of NOM*

Rajbhandari et al. (1999) and Wilczak et al. (2003) investigated the effect of treatment on subsequent chloramine decay. Both found that conventional treatment by coagulation, flocculation, and sedimentation increased chloramine stability compared to raw waters, and membrane nanofiltration (NF) improved chloramine stability even further. Wilczak et al. (2003) also studied the effect of ultrafiltration (UF) membranes and found them to increase chloramine stability.

With respect to ozonation and biofiltration, the study results conflicted. In the study by Rajbhandari et al. (1999), ozone/biofiltration, which decreases TOC, decreased chloramine demand as well. However, ozone/biofiltration increased chloramine demand in the Wilczak et al. (2003) study, whether sand or GAC media was used for biofiltration. Wilczak et al. (2003) showed that this increase in chloramine demand from ozone/biofiltration could be reduced by applying free chlorine or chlorine dioxide to biofilter effluent prior to chloramine formation.

In this study, new (non-biologically active) GAC and sand media filtration did not exhibit this negative effect on chloramine demand. Wilczak et al. (2003) did not determine whether sand or GAC filters were better suited to provide water that was more stable with respect to chloramine in the distribution system.

Both studies found that a higher initial chloramine dose resulted in slightly higher chloramine demand. Wilczak et al. (2003) found that a short free chlorine contact time had a stabilizing effect on subsequent chloramine decay in treated waters, while results by Rajbhandari et al. (1999) indicated that increased free chlorine contact time in treated waters did not significantly affect chloramine decay.

Duirk et al. (2001) investigated the fate and kinetics of monochloramine loss in the presence of NOM. Both a single and a dual reactive site model were used to model chloramine decay in the presence of fast reacting NOM (Suwannee River fulvic acid) and slow reacting NOM (Aldrich humic acid). UVA was measured and showed an overall trend in decreasing absorbance with monochloramine consumption. The UVA was measured over long periods of time, and other processes may have been occurring that could contribute to an overall change in absorbance (Duirk et al. 2001).

While studies suggest that the presence of NOM exerts an additional demand on monochloramine, especially at higher pH values, the research to date is inconclusive of the extent of the effects of NOM on monochloramine decay kinetics. Further investigation on chloramine behavior in the presence of NOM is necessary to gain a better understanding of the factors influencing chloramine decay kinetics with reactive waters.

## Chloramine Decay Models

Few models have been developed for chloramine decay in drinking water. Several studies have developed chloramine decay models in model solutions (Jafvert and Valentine 1992; Ozekin et al. 1996; Vikesland et al. 1996). In general, monochloramine decay includes many reactions that depend on several water quality parameters, including pH, ammonia concentration, ionic strength, and temperature. However, the above-cited models usually assume that no other oxidizable compounds besides ammonia are present in the water. Therefore, these models are not representative of real situations in water treatment in which NOM is present.

The chloramine decay algorithm in the USEPA WTP model (Version 1.0) was developed based on a study with sixteen bench scale tests conducted on waters obtained from utilities in the United States (MWDSC et al. 1991). The model, shown in Table 2.2, predicts chloramine decay using two first-order decay rate equations, one for the initial reaction up to 10 hours, and the second one from 10 to 180 hours. The chloramine residual decay experiments were conducted at 20°C, ambient pH, and chlorine to ammonia ratio of ≤4:1. The parameters $k_1$ and $k_2$ were reported to depend on the ratio of chlorine dose to TOC, ratio of UVA to TOC, ammonia concentration, turbidity, and alkalinity (Table 2.2).

**Table 2.2**
**Chloramine decay models**

| Study | Model | Parameters | Constraints |
|---|---|---|---|
| Harrington et al. 1992 (MWDSC et al. 1991) | for 0 < t < 10 hr<br>$\frac{[NH_2Cl]_t}{[NH_2Cl]_1} = \exp^{-k_1 t}$<br>for 10 < t < 180 hr<br>$\frac{[NH_2Cl]_t}{[NH_2Cl]_2} = \exp^{-k_2 t}$ | $[NH_2Cl]_1 = 0.439 + 0.460*\left[\frac{(Cl_2)_0}{TOC}\right] + 0.767*(Cl_2)_0 - 8.77*\left(\frac{UV_{254}}{TOC}\right)$<br>$[NH_2Cl]_2 = [NH_2Cl]_1 * \exp^{10*(k_2 - k_1)}$<br>For $\frac{(Cl_2)_0}{TOC} < 3.0\frac{\text{mg chlorine}}{\text{mg TOC}}$<br>$k_1 = 0.0452 - \left[0.00692*\frac{Cl_2}{TOC}\right] - \left[0.00661*\frac{(Cl_2)_0}{NH_3 - N}\right] + \left[0.199*\frac{UV_{254}}{TOC}\right] - 0.0000277(Alk)$<br>$k_2 = 0.0334 - 0.00240*\frac{Cl_2}{TOC} - 0.00664*\frac{(Cl_2)_0}{NH_3 - N} - 0.0000277*(Alk)$ | Limited data set.<br>T = 20°C.<br>Ambient pH.<br>Raw and coagulated waters. |
| Michaelis-Menten (Summers et al. 2001) | $\frac{dC}{dt} = \frac{k_1 C_0}{k_2 + C_t}$<br>Integrated form:<br>$C_t = \left[k_1 \cdot \ln\left(\frac{C_0}{C_t}\right)\right] - (k_2 \cdot t) + C_0$ | $C_t$ = $NH_2Cl$ concentration at time, t<br>$C_0$ = initial $NH_2Cl$ concentration<br>$k_1$ = - 0.977*$CA_0$ - 0.0352<br>$r^2 = 0.98$<br>$k_2$ = -0.0154 (UVA)<br>$r^2 = 0.79$ or<br>$k_2$ = -0.0185*UVA + 0.0006<br>$r^2 = 0.86$ | Limited data set of five waters.<br>Raw, settled and ozone/biofiltration waters.<br>T = 20°C.<br>pH = 8.0. |

Recently, a model was developed for chloramine decay based on the saturation type equation (Table 2.2) similar to that used by Koechling (1998) and Isabel et al. (2000) for chlorine decay (Summers et al. 2001). The model was developed for 52 experiments using five waters at different treatment levels: raw, settled, and ozonated/biofiltered. A sensitivity analysis showed that the predicted chloramine concentrations were most affected by changes in $k_1$ relative to changes in $k_2$. A 10 percent change in $k_1$ had the same effect on predicted chloramine concentrations as a 90 percent change in $k_2$. The rate parameter $k_1$ was found to be well correlated to the initial chloramine dose, $CA_0$, and the rate parameter $k_2$ was found to correlate well with initial UVA.

## CHLORINE DECAY KINETICS IN A SIMULATED DISTRIBUTION SYSTEM

Chlorine losses observed in distribution systems are due to reactions that occur within the bulk flow and also with material along the pipe. This phenomenon is illustrated in Figure 2.1. In this figure, free chlorine (HOCl) is shown reacting with NOM in the bulk phase to form DBPs and is also transported through a boundary layer at the pipe-wall to oxidize iron (Fe) released from pipe-wall corrosion. Other wall reactants might include biofilm and adsorbed organics.

The rates of bulk flow reactions ($K_{bulk}$) are affected primarily by the amount and nature of the organic matter remaining in the finished water. Because advanced treatment will change the nature or amount of TOC present, one expects that it will also affect the rates of chlorine and chloramine decay. Because these reactions are not affected by conditions at the pipe-wall, they can be studied and characterized independently at the bench-scale.

Pipe-wall reaction rates ($K_{wall}$) are affected by the nature of the material attached to or released from the wall as well as the rate at which chlorine or chloramine is transported to the wall reaction zone from the bulk fluid. In metal pipe, a major contributing factor to wall demand is the release of ferrous ions that will be oxidized to ferric by chlorine. Another demand for chlorine can be the organic matter sorbed onto ferric oxide or carbonate scales or contained in the exocellular polymeric material associated with biofilm growth on the pipe-wall. Advanced treatment that changes the amount and nature of the organics in water can affect the amount of wall-related organics available for reaction as well as the amount of available ferrous iron when iron-organic complexation is considered. In addition, changes in pH, alkalinity, and total dissolved solids can significantly affect the rate of corrosion-produced iron which in turn will affect chlorine reaction rates.

### Previous Work

Wable et al. (1991) were among the first to show that chlorine consumption within the pipes of a distribution system could be significantly higher than for the same water stored in a non-reactive glass bottle. Since then, a number of other studies have shown the same result. The overall decay of chlorine within a pipe is usually represented as a first-order reaction consisting of parallel first-order reactions occurring in the bulk flow and at the pipe-wall. The rate expression for this reaction can be written as shown in Equation 2.1.

$$\frac{dC}{dt} = -K_{total}C = -(K_{bulk} + K_{wall})C = -\left(K_{bulk} + \frac{S}{V}k_{wall}\right)C \qquad (2.1)$$

where $C$ = chlorine concentration (mass/volume) at time t
$K_{total}$ = overall reaction constant (1/time)
$K_{bulk}$ = bulk water reaction constant (1/time)
$K_{wall}$ = pipe-wall solution reaction constant (1/time)
$k_{wall}$ = pipe-wall surface reaction constant (length/time)
S/V = pipe-wall surface area per unit of pipe volume (1/length) For circular pipes, (S/V) is nominally equal to 4 divided by the pipe's diameter but might be higher in pipes with roughened surfaces due to encrustation and tuberculation.

Several studies have quantified the relative magnitude of the first-order rate constants for free chlorine decay in the pipe versus the bulk water alone. Table 2.3 compares the rate constants measured from several such efforts which were either

1. field studies measuring the input/output concentration of chlorine in a single length of pipe,
2. modeling studies where chlorine decay in a complete pipe network was calibrated to a set of geographically disperse measurements, or
3. experimental studies in pipeline simulators.

In this table, the wall surface rate constant ($k_{wall}$) is listed in addition to the solution rate constant ($K_{wall}$) so that measurements made on pipes of different diameters can be compared. Most of the metallic pipes have an overall decay constant that is 2 to 10 times higher than the bulk water rate constant.

More recently, Jones (2002) conducted a series of field studies on selected pipes of the distribution system serving the Norfolk (Virginia) Naval Base both prior to and after the water supplier switched from free chlorine to chloramines. Table 2.4 summarizes both the total and bulk reaction constants estimated from these data. The free chlorine constants are generally in the same range as those of other investigators shown in Table 2.3, while the numbers for chloramines are substantially lower.

Different studies have identified the various factors that affect the wall reaction rate for chlorine. These factors include pipe material, flow conditions, initial chlorine concentration, corrosion rate, and attached biomass.

- **Pipe Material**: Based on the work of Gotoh (1988), Sharp et al. (1991), and Jones (2002) it appears that the various pipe materials can be listed in order of increasing reactivity as follows: PVC, asbestos cement, cement-lined ductile iron, and unlined cast iron, with the latter material having substantially higher reactivity than the others. The reactivity of these materials is actually a result of other factors listed below.
- **Flow Conditions**: In laboratory pipe reactor tests, Kiene et al. (1993) observed almost negligible wall demand for chlorine under laminar flow as compared to turblent conditions, suggesting that mass transfer of chlorine to the pipe-wall could be a limiting factor in its reaction rate. Using a pipe simulator with Reynolds numbers in the turbulent flow range, Parent et al. (1996b) found that the reaction rate constant increased with increasing Reynolds number. Rossman et al. (1994) developed a model that incorporated mass transfer effects into the estimation of $k_{wall}$ and showed that this

**Table 2.3**
**Reaction rate constants for chlorine decay in pipelines**

| Investigators | Type of study | Pipe material | $K_{total}$ (1/day) | $K_{bulk}$ (1/day) | $K_{wall}$ (1/day) | $k_{wall}$ (ft/day) |
|---|---|---|---|---|---|---|
| Wable et al. (1991) | Field | Gray Cast Iron | 3.17 | 1.58 | 1.59 | 0.13 |
| | Field | Ductile Cast Iron | 9.94 | 2.88 | 7.06 | 0.58 |
| | Field | Ductile Cast Iron | 8.21 | 1.15 | 7.06 | 0.58 |
| Sharp et al. (1991) | Field | Unlined Cast Iron | 7.78 | N/A | N/A | N/A |
| | Field | Unlined Cast Iron | 5.76 | N/A | N/A | N/A |
| | Field | PVC | 4.90 | N/A | N/A | N/A |
| Kiene et al. (1993) | Field | Gray Cast Iron | 4.32 | 0.12 | 4.20 | 0.86 |
| | Field | Gray Cast Iron | 1.44 | 0.12 | 1.32 | 0.27 |
| Rossman et al. (1994) | Modeling | Unlined Cast Iron | N/A | 0.55 | N/A | 1.50 |
| Vasconcelos et al. (1997) | Modeling | Asbestos Cement | N/A | 1.16 | N/A | 0.00 |
| | Modeling | Unlined Cast Iron | N/A | 0.83 | N/A | 2.49 |
| | Modeling | Unlined Galvanized Iron | N/A | 0.23 | N/A | 0.89 |
| | Modeling | Unlined Cast Iron | N/A | 17.70 | N/A | 4.99 |
| | Modeling | Cement Lined Ductile Iron | N/A | 0.77 | N/A | 0.10 |
| Mathieu et al. (1993) | Simulator | Cement Lined Cast Iron | 2.90 | 0.00 | 2.90 | 0.24 |
| Parent et al. (1996b) | Simulator | Cement Lined Cast Iron | 3.60 | 0.52 | 3.08 | 0.25 |
| Parent et al. (1996a) | Simulator | Cement Lined Cast Iron | 12.10 | 0.54 | 11.56 | 0.98 |
| Rossman et al. (2001) | Simulator | Unlined Ductile Iron | 2.7-3.7 | 0.25-0.49 | 2.2-3.5 | 0.31-0.43 |

N/A indicates data not available

model was able to explain the variations of chlorine residuals measured in a distribution system.

- **Initial Chlorine Concentration**: Kiene et al. (1993) and Jones (2002) have noted that the wall reaction rate constant for chlorine increases with decreasing initial chlorine concentration. The usual explanation for this is that chlorine decay is not a true first-order reaction, and its rate is influenced by the concentration of other reactants besides chlorine.
- **Corrosion Rate**: Kiene and Levi (1995) used an electrochemical cell containing chlorinated water and an iron electrode to control the flux of $Fe^{+2}$. They found that rate of chlorine consumption was directly proportional to the rate of iron dissolution. Additional tests were made with water placed in a stirred annular reactor whose outer shell was cast iron pipe. Corrosivity of the water was modified using sodium chloride

**Table 2.4**
**Chlorine decay constants from pipes on the Norfolk Naval Base**

| Material | Free chlorine | | Chloramines | |
|---|---|---|---|---|
| | $K_{total}$ (1/day) | $K_{bulk}$ (1/day) | $K_{total}$ (1/day) | $K_{bulk}$ (1/day) |
| 12" PVC | 0.3–0.5 | 0.3–0.4 | 0.2–0.4 | 0.1–0.2 |
| 6" Ductile Iron | 0.6–1.4 | 0.6–1.4 | 0.24 | 0.08 |
| 8" Ductile Iron | 1.0 | 0.5 | N/A | N/A |
| 6" Asbestos-Cement | 2.1–4.0 | 0.5–1.6 | N/A | N/A |
| 10" Cast Iron | 1.3–5.1 | 0.5–1.5 | 0.4–1.0 | 0.13 |
| 8" Cast Iron | 3.1–12 | 1.0–4.6 | 0.5–2.2 | 0.12 |
| 8" Cast Iron | 3.5–5.1 | 0.9–1.3 | 0.9–3.8 | 0.05–0.5 |
| 6" Cast Iron | 4.9–7.4 | 0.6–2.4 | 1.1–8.1 | 0.05–0.4 |
| 6" Cast Iron | 0.8–4.4 | 0.4–2.1 | 0.1–0.7 | 0.1–0.2 |

and monitored using a Corrator™ probe. Results showed that chlorine decay proceeded at a constant rate with respect to time, which was proportional to the measured water corrosivity.

- **Attached Biomass**: Lu et al. (1999) studied the effect that biofilm growing on the pipe-wall surface has on chlorine demand. They found that the chlorine demand of biomass extracted from colonized glass beads was similar in strength to the bulk water demand. The rate constant for this demand was proportional to the biodegradable dissolved organic carbon (BDOC) in the water used to grow the biofilm.

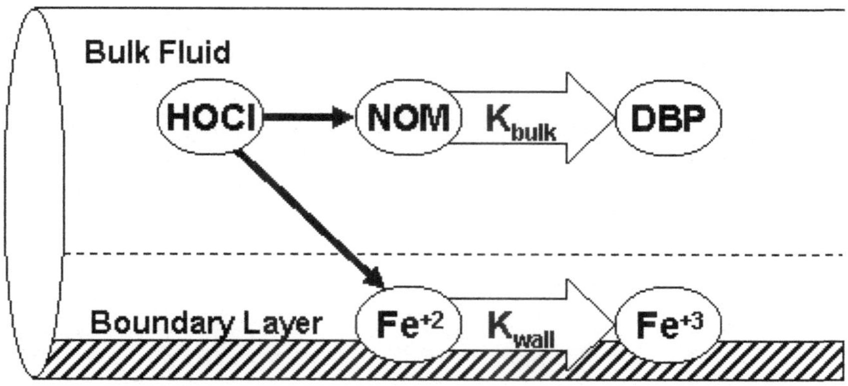


**Figure 2.1 Chlorine reactions occurring within a distribution system pipe**

# CHAPTER 3
# BENCH-SCALE TESTING

## OBJECTIVE

The main objective of the bench-scale studies was to determine the bulk chlorine and chloramine decay under conditions which are representative of practice, especially those of advanced treatment. To achieve this objective, three waters at different levels of treatment, disinfectant doses, temperature, pH, and blending conditions were evaluated. A secondary objective was to determine the effect of sediments from the distribution system on the decay of chlorine and chloramine.

## TESTING CONDITIONS

### Source Waters

Three source waters with varying water quality (TOC, SUVA, alkalinity, pH, and turbidity) were collected from three different WTPs:

Source #1) Betasso: Betasso Drinking Water Treatment Plant, Boulder, Colorado
Source #2) NPNV: Newport News Drinking Water Treatment Plant, Newport News, Virginia
Source #3) Winchester: Winchester Drinking Water Treatment Plant, Winchester, Kentucky

All source waters were collected and stored in 55-gallon precleaned plastic barrels. These barrels were acid/base washed and allowed to sit for 24 hours to leach organic material. They were then rinsed, dried, and sent to the sample location, where they were rinsed with sample water prior to sample collection. After arriving in the laboratory, the raw water was split into two portions. The portion of the raw water to be treated with bench-scale processes was enhanced coagulated, flocculated, settled, and filtered with a 1-µm cartridge filter. The remaining raw water was filtered with a 1-µm cartridge filter to prepare it for the chlorination tests.

### Base Case Conditions

The disinfectant decay studies were conducted using the following base case conditions that are reflective of average operation conditions:

- temperature: 20°C
- pH: 8.0 (buffered)
- chlorine dose: 1.5:1 $Cl_2$:TOC
- chloramine dose: 4.5:1 $Cl_2$:N and target residual of 2 mg/L as $Cl_2$ at 24 hr.

For base case chlorine decay experiments, a defined $Cl_2$:TOC dose ratio was chosen over a defined residual. This provided a better base for comparing the reactivity of different NOM characteristics, opposed to quantity of organic carbon, between different source and treated waters. Chloramine decay is much slower than chlorine decay, and the majority of chloramine decay is due to auto-decomposition, as opposed to reactions with NOM. Therefore, the base case chloramine experiments were dosed to achieve a consistent chlorine residual after 24 hours, and ammonia was added accordingly to provide a $Cl_2$:N ratio of 4.5, considered optimal for avoiding dichloramine formation while limiting excess ammonia.

### Treatment Technologies

All of the waters were treated in a batch-mixing tank by enhanced coagulation, tapered flocculation, and settling. After enhanced coagulation, the waters were then treated using the following advanced treatment processes: GAC adsorption, ozone/biofiltration, ozone only, and UV irradiation. The following abbreviations will be used to identify the treatment technologies:

- *Raw:* untreated water
- *EC:* enhanced coagulation, flocculation, and settling followed by a 1 μm filter
- *GAC:* GAC adsorption of EC water
- $O_3$ *Only:* ozonation of EC water
- $O_3/BF$: ozonation followed by biofiltration of EC water
- *UV:* UV irradiation treatment of EC treated water

After each treatment level is a chlorination/chloramination point. Figure 3.1 is a flow diagram showing the different treatment levels and all the points where chlorination or chloramination experiments were performed.

### Disinfectant Dose

#### *Chlorine*

All of the waters (treated and untreated) were dosed at the base case dose of 1.5:1 $Cl_2$:TOC. Chlorine decay experiments at an additional higher dose ratio (2.5:1 $Cl_2$:TOC) were performed on all waters from Source #1 and most waters from Source #2. The remaining waters from Source #2 were dosed at an additional lower dose ratio (0.75:1 $Cl_2$:TOC). These dose ratios cover a range representative of disinfectant residuals found in practice. For the experiments performed at the additional doses, all other reaction conditions were held at the base case. After analyzing the data from the first two waters, it was determined that chlorine kinetics could be predicted for other dose ratios based on a single chlorine decay experiment. Therefore, for all of the treated waters from Source #3, experiments were performed only at the base case dose. The single exception to this is that the EC water was also dosed at the lower dose ratio.

#### *Chloramine*

All of the waters (treated and untreated) were dosed at a 4.5:1 $Cl_2$:N ratio. Both a low dose and a high dose chloramine decay study were conducted on all of the waters. The low dose targeted a 24-hr. chloramine residual of 2 mg/L as $Cl_2$. The high dose targeted a 24-hr. chloramine residual of 4 mg/L as $Cl_2$.

## Temperature

In addition to the base case temperature (20°C), EC treated waters from all three sources were chlorinated and chloraminated at temperature conditions of 30°C and 4°C. Water from one advanced treatment condition from each source water was chloraminated at temperature conditions of 30°C and 4°C.

## pH

### *Chlorine*

The effect of pH was evaluated on all three waters after EC treatment. In addition to the base case pH of 8.0, the chlorine decay at pH 5.0 and pH 11.5 were evaluated with all other conditions set at the base case. The effect of changing pH during chlorination was also examined on two EC source waters. Both waters were chlorinated at the base case conditions except for pH. The pH was varied with time in the pH range of 5 to 9. This situation was set in a manner to simulate full-scale behavior.

### *Chloramine*

The effect of high pH was evaluated on all three waters after EC treatment. In addition to the base case pH of 8.0, a chloramine decay study at pH 10.5 was conducted with all other conditions set at the base case.

## Decay Curve Sampling Interval

The raw water and each of the treated waters were dosed with chlorine and analyzed for chlorine residual following these specified times: 2.5, 5, 10, 15, 30, 45, and 60 minutes, and 1.5, 2, 3, 5, 8, 24, 72, and 120 hours. At the end of each chlorine decay study (120 hours), the pH was measured to ensure that it had stayed within 0.2 pH units of the target pH.

The chloramine residuals of the raw water and each of the treated waters was analyzed following these time frames: 15, 30, and 45 minutes, and 1, 2, 4, 8, 24, 72, and 120 hours. Chloramine residual readings extended further to include 10 and 20 days.

UVA at 254 nm ($UVA_{254}$) was monitored in conjunction with the decay of chlorine and chloramine. For both chlorine and chloramine decay studies, the $UVA_{254}$ was measured at the same reaction times as chlorine residual.

## Water Quality Parameters

Table 3.1 lists the water quality parameters that were measured for each of the source waters. Samples were analyzed in duplicate for quality assurance, and the two results were averaged.

**Table 3.1**
**Bench-scale water quality parameters**

| | TOC | DOC | $UVA_{254}$ | pH | Alkalinity | Hardness | Turbidity | Ammonia | Bromide |
|---|---|---|---|---|---|---|---|---|---|
| Raw | ✓ | ✓ | ✓ | ✓ | ✓ | ✓ | ✓ | ✓ | ✓ |
| EC | ✓ | | ✓ | ✓ | ✓ | ✓ | ✓ | ✓ | |
| GAC | ✓ | | ✓ | ✓ | | | | | |
| $O_3$/BF | ✓ | | ✓ | ✓ | | | | | ✓ |
| $O_3$ Only | ✓ | | ✓ | ✓ | | | | | ✓ |
| UV | ✓ | | ✓ | ✓ | | | | | |

**Table 3.2**
**Enhanced coagulation requirements of the D/DBP Rule for percent TOC removal**

| Initial TOC (mg/L) | Required TOC removal (percent) alkalinity (mg/L as $CaCO_3$) | | |
|---|---|---|---|
| | <60 | 60–120 | >120 |
| >2-4 | 35% | 25% | 15% |
| >4-8 | 45% | 35% | 25% |
| >8 | 50% | 40% | 30% |

## EXPERIMENTAL DESIGN AND EQUIPMENT

### Enhanced Coagulation

Traditionally, coagulation has been used primarily for turbidity removal, contributing to the control of microbiological contaminants such as viruses, bacteria, and protozoa. However, Stage 1 of the D/DBP Rule includes an enhanced coagulation TOC removal requirement. This requires conventional treatment plants to remove NOM to a specified degree, depending on the TOC and alkalinity of the source water as shown in Table 3.2.

Jar tests were performed on each of the three source waters to determine the appropriate dose of alum coagulant ($Al_2(SO_4)_3$) needed to meet the enhanced coagulation requirements. Six rectangular jars (Phipps & Bird B-KER$^{2®}$ Laboratory Jars) were filled with the source water under investigation. The jars hold 2 L of water and have sample ports 5 cm, or a third of the depth, above the bottom. The jars were mixed at the mixing speeds and duration shown in Table 3.3 using a Phipps & Bird PB-900 Programmable Jar Tester, which stirs at four sequential memory speeds. A schematic of the jar test mixing sequence is shown in Figure 3.2.

Different volumes of alum were added to each of the six jar test containers, in a range representative of water treatment practice. The TOC, pH, and turbidity of each of the finished jar

**Table 3.3**
**Jar test and EC mixing conditions**

| | Rapid mix | Tapered | Flocculation | Sedimentation |
|---|---|---|---|---|
| Mixing speed (rpm) | 100 | 30 | 10 | None |
| Duration (minutes) | 3 | 15 | 15 | 60 Jar Test, 180 EC |

tested waters were analyzed. From these results, the appropriate alum dose was selected in order to comply with enhanced coagulation requirements of the D/DBP Rule (Table 3.2).

After jar testing was complete and an appropriate alum dose was selected, a large (40 L) mixing tank with scaled-up dimensions of the individual jar testers was used to produce large quantities of EC treated water for chlorination/chloramination experiments and further treatment. The waters were mixed in the 40L tanks using the same mixing conditions as the jar tests (Table 3.3, Figure 3.2). The coagulated and flocculated water was allowed to settle at least five hours in the larger tank. After settling, the water was collected from the tank and filtered by a 1-μm cartridge filter.

## Advanced Treatment Processes

All of the advanced treatment processes (GAC, $O_3$ Only, $O_3$/BF, and UV) were performed on enhanced coagulated waters. Following the designated advanced treatment of EC water, water quality data were collected, and the newly treated water was stored at 4°C. Both chlorine and chloramine decay experiments were conducted within 30 days of treatment.

### *GAC Adsorption*

For the water requiring GAC treatment, EC water was recirculated through a bench-scale GAC column containing 100 mg of GAC per liter of water in the reservoir. A schematic of the bench-scale system is shown in Figure 3.3. The flow into the GAC column was high enough to yield no measurable concentration gradient, thus yielding a differential column reactor (Summers 1986). TOC of the recirculating water was measured until 25 percent TOC removal was achieved, which is the upper end of removals found in practice. The mixed reservoir water in the batch reactor was recirculated through the column two to three times to achieve this removal.

### *Ozonation*

For both $O_3$ Only and $O_3$/BF, an ozone solution was made by injecting ultra-pure water with ozone to a desired concentration of approximately 25 mg/L $O_3$. An Ozone Research & Equipment Corporation (OREC) batch ozonator (Model O3V5-0) was used to make the ozone stock solution. The exact concentration of the batch ozone solution was determined before the samples were dosed. The UV absorbance at 258 nm was measured per centimeter. From this, the ozone concentration in mg/L was determined using Equation 3.1.

$$O_3\ Batch\ Strength = \frac{UVA_{258}}{3000M^{-1} \cdot cm^{-1}} \cdot \left(48{,}000 \frac{mg}{mol} O_3\right) \tag{3.1}$$

While keeping the sample water on a stir plate to ensure thorough mixing, the ozone solution was then directly added to the sample waters using a 100-mL syringe. Once the appropriate amount of ozone solution was added to the sample water, the newly ozonated sample water was left to react until all of the ozone had dissipated (approximately 5 hours).

For $O_3$ Only water, chlorine/chloramine decay studies were conducted immediately following the complete dissipation of ozone. For $O_3$/BF water, the ozonated water was passed through a biofiltration reactor.

***Biofiltration***

A bench-scale recirculating biofiltration reactor was used for ozonated water to complete the $O_3$/BF process. A schematic of this system, which is similar to the GAC system, is shown in Figure 3.3. The water was recycled from a mixed reservoir through biologically acclimated sand and then returned to the reservoir. The water in the reactor reservoir was monitored twice daily for DOC, and after fifteen to twenty percent of the DOC (the fast reacting BDOC) had been biologically reacted, the water was separated from the biologically acclimated sand media. When treating NPNV, the water was left in the reactor too long due to a problem with the TOC analyzer, leading to a 26% TOC removal via biofiltration.

The sand column was 1.0 inch in diameter and 10.6 inches deep and the water was recirculated over the sand at a loading rate of 1.9 gpm/ft$^2$. In order to achieve fifteen percent TOC removal, the water needed to be recirculated through the reactor between two and three times (empty bed contact times between 7 and 10 minutes).

***UV Irradiation***

After the water had been EC treated, the designated volume was passed through a bench-scale, low-pressure, single-pass flow-through unit (SP-1 Aquafine Ultraviolet Disinfection Unit). In this unit, the water flows through a clear quartz tube with a long UV lamp parallel to it. The UV lamp (Aquafine Corp. 3010) is a low-pressure mercury discharge lamp, meaning that it produces nearly monochromatic light at 254 nm. In addition, there is a quartz lamp envelope material that only allows UV energy wavelengths of 254 nm to be transmitted to the water stream. UV dose was calculated according to Equation 3.2.

$$D \times I \times t \tag{3.2}$$

where D = UV Dose (mJ/cm$^2$)
I = Intensity (mW/cm$^2$)
t = time (sec)

A contact time of 3.7 seconds was achieved, calculated according to Equation 3.3.

$$t = \frac{V_T}{Q} \tag{3.3}$$

where $V_T$ = volume of the quartz tube (measured at $1.2 * 10^{-4}$ m$^3$)
Q = flow through the tube (0.12 m$^3$/hr)

The intensity of the UV lamp was 11 mW/cm$^2$. Therefore, using Equation 3.2, the approximate applied UV dose was 41 mJ/cm$^2$. This dose is sufficient to achieve 4-log inactivation of *Cryptosporidium* (Clancy et al. 2000).

### Pipe Sediments

A small (one-foot) section of a 6-inch unlined ductile iron pipe was taken from the City of Boulder's drinking water distribution system when a section of pipe was being replaced. The pipe was immediately placed in tap water at 4°C, so the sediments lining the pipe remained acclimated to treated, chlorinated water. A wire brush was used to scrape some of the iron sediment from the pipe. This was collected and also held in tap water at 4°C until it was used in chlorine decay experiments. Before a sediment experiment was performed, a portion of wet sediment was weighed out to achieve the desired wet sediment concentration.

A volume of 19.5 L of the water was measured out and added into three 20-L carboys. Equal amounts of chlorine were added to each carboy. The chlorine residual of each solution was measured after three minutes to ensure they had been dosed the same. After five minutes, sediment was added. The first carboy was a control experiment, in which no sediment was added. The second and third carboys received 0.5 and 1.0 grams of wet sediment per liter.

### Chlorination, Chlorine Decay, and UVA$_{254}$ Decay

For each of the chlorination decay experiments, the water was dosed in a one-gallon amber bottle. First, the bottle was filled three-quarters full with sample water, and the appropriate chlorine dose was added with a pipette. The method for measuring the strength of the chlorine dosing solution to determine the appropriate dose is described in the analytical methods section. After dosing the water, the cap was placed on the bottle, and the bottle was gently inverted twice to mix the chlorine with the water. Lastly, the cap was removed, the bottle was filled the rest of the way with sample water, and the cap was replaced on the bottle headspace-free.

The moment the chlorine dose was added to the sample was considered time 0 for the decay experiments. Chlorine residual and UVA$_{254}$ were measured after 2.5, 5, 10, 15, 30, 45, and 60 minutes; and 1.5, 2, 3, 4, 8, 24, 72, and 120 hours.

For the first fifteen minutes after chlorination, the samples were taken directly from the one-gallon bottle for analysis. At the appropriate time for a chlorine residual measurement, 100 mL were poured into a 100-mL graduated cylinder and analyzed for chlorine residual. Also, 3.5 mL were poured into a 10-mm quartz cell (cuvette) and analyzed for UVA$_{254}$.

After the first four chlorine residual and UVA$_{254}$ analyses were taken, the remaining chlorinated water was poured, headspace free, into individual 125-mL amber bottles. These were stored in the dark at 20°C unless otherwise stated. At the appropriate time for a chlorine residual measurement, 100 mL from a 125-mL amber bottle was poured into a graduated cylinder for a chlorine residual reading, while some of the remaining water (approximately 5 mL) was poured into a cuvette and analyzed for $UVA_{254}$. At the end of each of the chlorine decay experiments (120 hours), the remaining 20 mL of the last sample was used to measure pH.

### Chloramination, Chloramine Decay, and $UVA_{254}$ Decay

For each chloramination decay experiment, the water was dosed in a 0.7-gallon amber bottle. First, the 0.7-gallon amber bottle was filled with 2.5 L of sample water and placed on a stir plate to ensure thorough mixing. Then, the appropriate volume of ammonium sulfate solution was added with a pipette. After the addition of ammonium sulfate, the appropriate volume of chlorine dosing solution was added to yield a 4.5:1 $Cl_2$:N ratio.

The timer was started simultaneously with the addition of the chlorine dosing solution. A chloramine residual measurement and $UVA_{254}$ analysis was taken 15, 30, and 60 minutes; 1, 2, 4, 8, 72, and 120 hours; and 10 and 20 days.

For the first thirty minutes after chloramination, the samples were taken directly from the 0.7-gallon bottle for analysis. At the appropriate time for a chloramine residual measurement, 100 mL was poured into a 100-mL graduated cylinder and analyzed for chlorine residual. Also, 3.5 mL were poured into a 10-mm cuvette and analyzed for $UVA_{254}$.

After 30 minutes, the remaining chloraminated water was poured, headspace free, into individual 125-mL amber bottles. These were stored in the dark at 20°C unless otherwise stated. At the appropriate time for a chloramine residual measurement, 100 mL from a 125-mL amber bottle was poured into a 100-mL graduated cylinder for a chloramine residual reading, while the remaining water was poured into a cuvette and analyzed for $UVA_{254}$.

### Temperature

For the analysis of temperature effect, refrigerators or water baths held the temperature of the samples constant at 4°C, 20°C, or 30°C (±2°C).

### pH

Chlorination/Chloramination was performed at pH 8.0 (±0.2) unless otherwise stated. Before dosing, the sample water was buffered to pH 8.0 (±0.2) with approximately 2 mL/L borate buffer: 1.0 M boric acid (American Chemical Society [ACS] grade) and 0.26 M sodium hydroxide (ACS grade) in ultra-pure water (Summers et al. 1996). The sodium hypochlorite dosing solution was made from a 10.5 percent sodium hypochlorite stock solution and buffered to pH 8 with pH 6.7 borate buffer (1.0 M boric acid (ACS grade) and 0.11 M sodium hydroxide (ACS grade) in ultra-pure water (Summers et al. 1996).

## ANALYTICAL METHODS

### Summary of Analytical Methods and Sampling Procedure

Table 3.4 is a summary of the analytical methods used in this study, and Table 3.5 contains a summary of the collection and preservation of samples for analytes requiring special handling.

**Table 3.4**
**Bench-scale analytical methods**

| Analyte | Measuring units | Method detection limit | Equipment/procedure | Reference method* |
|---|---|---|---|---|
| TOC/DOC | mg/L | 0.05 mg/L | Sievers 800 Total Organic Carbon Analyzer with Auto-sampler | SM 5310 C |
| UVA | $cm^{-1}$ | 0.001 $cm^{-1}$ | Hach DR 4000 UV-Visible Recording Spectro-photometer | SM 5910 B |
| pH | N/A | N/A | Corning Model 430 meter and probe | SM 4500-$H^+$ |
| Hardness | mg/L as $CaCO_3$ | 2 mg/L as $CaCO_3$ | Hach Digital Titrator Model 16900-01 | SM 2340 C |
| Alkalinity | mg/L as $CaCO_3$ | 2 mg/L as $CaCO_3$ | Hach Digital Titrator Model 16900-01 | SM 2320 B |
| Turbidity | NTU | 0.001 NTU | Hach 2100N Turbidimeter | SM 2130 B |
| Bromide | µg/L | 4 µg/L | Dionex DX-300 Series Chromatography System | EPA Method 300 B |
| Chlorine strength | mg/mL | 40 µg/L | Iodometric Method I | SM 4500-Cl B |
| Chlorine residual | mg/L as $Cl_2$ | 18 µg/L as $Cl_2$ | DPD Ferrous Titrimetric Method | SM 4500-Cl F |
| Chloramine residual | mg/L as $Cl_2$ | 18 µg/L as $Cl_2$ | DPD Ferrous Titrimetric Method | SM 4500-Cl F |

* SM = Standard Methods

Although Table 3.5 includes maximum storage times, all analytes were measured as quickly as possible, and all preservation techniques were strictly followed.

## Data Quality Acceptance Criteria

After arrival, each raw water source was sampled in duplicate for TOC, DOC, $UVA_{254}$, bromide, pH, turbidity, alkalinity, and hardness. Water was stored in a refrigerator at 4°C until treatment and chlorination. After each experimental treatment process, pH, TOC, and $UVA_{254}$ were analyzed in duplicate. In addition, alkalinity, hardness, and turbidity were analyzed in duplicate after EC treatment, and bromide was analyzed in duplicate after $O_3$/BF treatment. The average value for each set of duplicate measurements is reported. For each kinetic decay experiment, four of the chlorine and chloramine residual samples had a qualitative, duplicate sample in a separate bottle from the original.

**Table 3.5**
**Summary of sampling procedures (APHA, AWWA, and WEF 1998)**

| Analyte | Minimum volume (mL) | Container required | Preservation | Maximum storage |
|---|---|---|---|---|
| pH | 50 | P, G | Analyze immediately | 0.25 hr |
| TOC / DOC | 100 | G (B) | Analyze immediately or refrigerate and add $H_3PO_4$ to pH < 2 | 7 d |
| Hardness | 100 | P, G | Add $HNO_3$ or $H_2SO_4$ to pH < 2 | 6 months |
| Alkalinity | 200 | P, G | Refrigerate | 24 hr |
| Turbidity | 100 | P, G | Analyze same day, store in dark up to 24 hrs, refrigerate | 24 hr |
| Bromide | 100 | P, G | None required | 28 d |
| Chlorine | 100 | P, G | Analyze immediately | 0.25 hr |
| Chloramine | 100 | P, G | Analyze immediately | 0.25 hr |

Minimum volume based on Standard Methods recommendation. In actuality, may be smaller than indicated value because experimental equipment requires less volume.

P = plastic (polyethylene or equivalent); G = glass; G (B) = glass, borosilicate; poly tetra fluoroethlyene (PTFE). Acid washed.

There is no special glassware recommendation for chlorine analysis. However, amber, chlorine demand free glassware was used.

## RESULTS

### Water Quality

Tables 3.6, 3.7, and 3.8 show the water quality before and after treatment from Betasso, NPNV, and Winchester source waters, respectively. NPNV source water had the highest TOC concentration, whereas both Betasso and Winchester had more moderate TOC concentrations. All three source waters had moderate TSUVA (UVA/TOC), low alkalinity, and low hardness. Betasso and NPNV contained low source water turbidity, and Winchester source water turbidity was high.

EC followed by filtration removed over fifty percent of the TOC for Betasso and NPNV waters and twenty percent for the Winchester water. EC lowered the TSUVA for the first two source waters, indicating that coagulation preferentially removed organic matter containing carbon-carbon double bonds and aromatics often associated with the hydrophobic fraction. TSUVA of the Winchester water did not change significantly after EC treatment. EC treatment significantly lowered the turbidity, alkalinity, and pH of all three waters. UV treatment of EC water did not affect the TOC for any of the waters, but it did lower the TSUVA for NPNV water, indicating a possible reaction with the carbon-carbon double bonds. The TSUVA was not decreased after UV treatment for the other two waters. GAC treatment of the EC waters was performed to remove around twenty-five percent of the TOC. It lowered the TSUVA, indicating a preferential removal of organic matter containing carbon-carbon double bonds and aromatics. Water quality after ozonation was measured for NPNV and Winchester waters. In both cases the

**Table 3.6**
**Water quality parameters and effect of treatment for Betasso water**

| Treatment level | TOC (mg/L) | UVA ($cm^{-1}$) | TSUVA (mg/ L·m) | Turbidity (NTU) | Alkalinity (mg/L as $CaCO_3$) | Hardness (mg/L as $CaCO_3$) |
|---|---|---|---|---|---|---|
| Raw | 3.24 | 0.096 | 2.97 | 2.1 | 12 | 17 |
| EC | 1.52 | 0.028 | 1.85 | 0.15 | 6.5 | — |
| UV | 1.69 | 0.032 | 1.90 | — | — | — |
| GAC | 1.15 | 0.020 | 1.74 | — | — | — |
| $O_3$/BF | 1.10 | 0.014 | 1.27 | — | — | — |

**Table 3.7**
**Water quality parameters and effect of treatment for NPNV water**

| Treatment level | TOC (mg/L) | UVA ($cm^{-1}$) | TSUVA (mg/ L·m) | Turbidity (NTU) | Alkalinity (mg/L as $CaCO_3$) | Hardness (mg/L as $CaCO_3$) |
|---|---|---|---|---|---|---|
| Raw | 5.52 | 0.208 | 2.51 | 1.3 | 33 | 43 |
| EC | 3.56 | 0.068 | 1.89 | 0.28 | 20 | — |
| UV | 3.53 | 0.059 | 1.67 | — | — | — |
| GAC | 2.59 | 0.042 | 1.62 | — | — | — |
| $O_3$ | 2.84 | 0.015 | 0.53 | — | — | — |
| $O_3$/BF | 2.11 | 0.021 | 1.00 | — | — | — |

**Table 3.8**
**Water quality parameters and effect of treatment for Winchester water**

| Treatment level | TOC (mg/L) | UVA ($cm^{-1}$) | TSUVA (mg/L·m) | Turbidity (NTU) | Alkalinity (mg/L as $CaCO_3$) | Hardness (mg/L as $CaCO_3$) |
|---|---|---|---|---|---|---|
| Raw | 2.44 | 0.054 | 2.21 | 100 | 49 | 48 |
| EC | 1.94 | 0.043 | 2.22 | 0.15 | 42 | — |
| UV | 1.94 | 0.050 | 2.58 | — | — | — |
| GAC | 1.30 | 0.027 | 2.04 | — | — | — |
| $O_3$ | 1.71 | 0.014 | 0.82 | — | — | — |
| $O_3$/BF | 1.42 | 0.016 | 1.13 | — | — | — |

TOC was reduced and the TSUVA was significantly lowered, likely due to the oxidation of carbon-carbon double bonds. Biofiltration was performed targeting an additional twenty-five percent TOC removal. This process increased the TSUVA after ozonation, indicating a preferential removal of non-aromatic organic matter. However, biofiltration did not restore the TSUVA to the level it had been before ozonation.

## Chlorine Decay Primary Results

The primary results of this study are presented in terms of chlorine residual concentration versus time. In this section, the effects of the different variables on these results are discussed. For many of the scenarios, additional data analysis is provided in the next section. All primary data are reported in Appendix A. After chlorination, there is an almost instantaneous chlorine demand in drinking water, thought to be largely due to oxidation of reduced inorganic compounds and some very easily oxidized organic compounds. After the initial demand, the chlorine reacts quickly during the first few hours. This phase takes place predominantly in the treatment plant. The short-term decay is followed by a slow long-term decay which dominates in the distribution system. The short-term and long-term chlorine decay are due to chlorine reactions with different NOM species.

Figure 3.4 shows the decay of chlorine in EC NPNV water chlorinated at three different doses under base case conditions (pH 8.0, and 20°C). These decay curves are typical, exhibiting the three phases of chlorine decay discussed previously. To visualize the decay in the first few hours more easily, the time axis can be plotted logarithmically (Figure 3.5).

### *Effect of Dose*

For all three doses, short-term decay is very fast compared to the subsequent slow, steady long-term decay. The higher chlorine dose yields a higher residual but also results in a higher chlorine demand at each point in time. The table in Figure 3.5 shows the 24-hour chlorine demand ($CD_{24}$) for each dose. This effect of chlorine dose on chlorine decay was seen at all three treatment levels for the Betasso and NPNV waters and for Winchester EC.

### *Effect of pH*

To investigate the effect of pH on chlorine decay kinetics, EC waters were buffered to pH 5.5, 8.0, and 10.5 and then chlorinated. A consistent pattern was not found for the short-term period. For Betasso (Figure 3.6), the high pH (10.5) displayed the slowest short-term decay. For NPNV (Figure 3.7), the low pH (pH 5.5) water reacted the slowest in the short term. For Winchester (not shown), the high pH and middle pH (10.5 and 8.0, respectively) reacted the same in the short-term period, which was slower than at pH 5.5. For all three waters, the chlorine residuals in the long-term time periods were similar, with the low pH (5.5) being the most stable. For Betasso and NPNV waters, the solutions at pH 8.0 and pH 10.5 exerted the same 120-hour chlorine demand ($CD_{120}$), while at pH 5.5 they had a smaller $CD_{120}$. Koechling (1998) found similarly unsystematic effects of pH on chlorine decay.

The pH of a water can change during treatment. In a conventional plant, alum coagulation can lower the pH to below 6. Then, prior to discharge to the distribution system, the pH is often raised to pH 8 or higher. Chlorine decay studies were performed on Betasso and NPNV EC waters to simulate this pH change. Each water was buffered to 5.5 for the first five hours of chlorine contact time to simulate alum coagulation optimized for enhanced coagulation. After five hours, the water was adjusted to pH 8.0 to simulate pH in the distribution system. For both waters, the chlorine concentration appeared to follow constant pH 8.0 decay after the pH was adjusted, with no apparent effect of the first five hours at low pH. Figure 3.8 shows the results from this study for EC NPNV.

### *Effect of Temperature*

To investigate the effect of temperature on chlorine decay kinetics, chlorine decay studies were performed at 4, 20, and 30°C. The effect of temperature on chlorine decay for Betasso is shown in Figure 3.9. NPNV and Winchester results were similar. For all three waters, the long-term chlorine decay was faster at higher temperatures while the chlorine decay in the first three hours was not significantly affected by temperature. In Betasso water, no effect was observed during the first three hours. In NPNV water, the chlorine residuals differed by less than 4 percent at three hours. In Winchester water, the 4°C and 20°C waters exhibited the same chlorine decay kinetics up to three hours, during which the chlorine decayed slightly faster in the 30°C water over the short-term period. The difference in three-hour residuals was approximately 20 percent. The effect of temperature is discussed further in the chlorine decay modeling section.

### *Effect of Treatment*

The effect of treatment on chlorine decay for Betasso, NPNV, and Winchester source waters are shown in Figures 3.10 through 3.12. All waters were dosed at a 1.5:1 $Cl_2$:TOC ratio, so the higher TOC waters were dosed at higher chlorine concentrations. However, these different doses yielded similar chlorine residuals at 120 hours. Thus, the $CD_{120}$ values were greater for the higher TOC waters. However, the specific chlorine demands (chlorine demand in mg per mg of TOC or $CD_{120}$/TOC) were similar (between 1.1 and 1.5 mg/mg) for all treatment levels for all three waters. Similar observations were made at the other dose ratios. The effect of treatment on chlorine decay is described further in the data analysis section.

### *Effect of Blending Chlorinated Waters*

In a drinking water distribution system, pipes are often laid out in a complex matrix where water straight from the treatment plant can mix with water that has been out of the plant for several days. The two waters would have different chlorine residuals and NOM sites that react differently with chlorine. Two blending experiments were performed with the Betasso GAC treated water to determine the effect on chlorine decay. In the first experiment (Figure 3.13), water that had been in contact with chlorine for 24 hours (termed "old" water) was blended with water that had been in contact with chlorine for only 2 hours (termed "new" water) at a 1:1 volume ratio. Immediately after blending, the combined water had a chlorine residual in between the new and old waters, but after two days the blended water residual was higher than that of the new water. Because 50 percent of the blended water NOM had reacted with chlorine for 24 hours, the blended water was more stable than the new water. The additional chlorine from the new water acted like rechlorination to the old water. In a second blending scenario, where the new water (2-hour contact) was reacted with 5-day old water, the results were similar.

### *Effect of Rechlorination*

After the initial blending experiments showed an effect similar to rechlorination of the old water, it was decided that additional rechlorination experiments would be conducted. Rechlorination experiments were carried out on all three source waters at the EC treatment level. The results from NPNV are shown in Figure 3.14. Three different experiments are shown in which chlorine was added to the water for a total dose of 1.5:1 $Cl_2$:TOC (mg/L). The first experiment is a base case, single-chlorination experiment. The other two experiments were both rechlorination experiments in which the water was dosed twice. The first water was rechlorinated after 24 hours and the second water after 72 hours. Both waters were dosed initially to target a residual of 0.3 mg/L just before rechlorination. The 24-hour rechlorinated water had the same residual as the base case water three days after chlorination, and the 72-hour rechlorinated water had the same residual as the base case water two days after rechlorination. It appears that the same net dose eventually results in the same chlorine residual regardless of rechlorination. Similar results were found with the other two waters.

### *Effect of Ductile Iron Pipe Sediments*

The effect of different masses of ductile iron pipe sediment on chlorine decay was investigated using tap water from the City of Boulder's drinking water distribution system. The results for the study are shown in Figure 3.15. Three different experiments were conducted. The first was a control experiment with no iron pipe sediments added to the water. The second and third experiments added 0.5 g/L and 1.0 g/L of iron pipe sediments to the test water. In these tests, the sediment was finely ground such that the suspension did not completely settle during the five-day experiment. The chlorine residual of the tap water was measured as 0.7 mg/L. The water was rechlorinated at time zero with 2.0 mg/L $Cl_2$. Initial chlorine residual measurements were taken after three minutes to ensure that the experiments were dosed the same, and the sediment was added to the reactors at five minutes. The ground sediment had a large effect on chlorine decay, greatly increasing the kinetics. As can be seen in Figure 3.15, the ground sediments change the shape of the chlorine decay curve.

In a second sediment experiment, large particles of sediment were placed in the chlorinated water. The doses of these non-ground sediments were 0.5 and 1.0 g/L, the same as the doses of the ground particles in the first tests. In this scenario, the large sediment had very little effect on chlorine decay, with the difference of chlorine residuals after 24 hours being less than 0.2 mg/L for water containing no sediment and the solutions with the two doses of sediment.

## Chlorine Decay Data Analysis

### *Chlorine Demand*

Table 3.9 shows chlorine demands at times of 1 hour, 24 hours, and 120 hours for all base case chlorine decay experiments at all treatment levels and doses. Also in this table are percentages of the long-term chlorine demands ($CD_t/CD_{120}$ * 100%) and specific chlorine demands ($CD_t$/TOC).

For all base case chlorine decay experiments in which a residual was present after 120 hours, between 25 and 50 percent (avg. = 36 percent) of the 120-hour chlorine demand was

**Table 3.9**
**Bench-scale chlorine demand results (pH 8.0; 20°C)**

| Source water | $Cl_2$:TOC dose (mg/mg) | Treatment level | TOC (mg/L) | Chlorine demand $CD_t$ (mg/L) | | | Percent demand $CD_t/CD_{120}$ (mg/mg) | | Specific demand $CD_t$:TOC (mg/mg) | | |
|---|---|---|---|---|---|---|---|---|---|---|---|
| | | | | 1-hr | 24-hr | 120-hr | 1-hr | 24-hr | 1-hr | 24-hr | 120-hr |
| NPNV | 0.75 | EC | 3.56 | 1.0 | 2.0 | | | | 0.3 | 0.6 | |
| NPNV | 0.75 | $O_3$ | 2.84 | 1.0 | 2.1 | | | | 0.4 | 0.7 | |
| NPNV | 0.75 | $O_3$/BF | 2.11 | 0.5 | 1.2 | | | | 0.2 | 0.6 | |
| Winchester | 0.75 | EC | 1.94 | 0.7 | 1.2 | | | | 0.3 | 0.6 | |
| Betasso | 1.5 | Raw | 2.86 | 1.5 | 2.9 | 4.1 | 37 | 72 | 0.5 | 1.0 | 1.4 |
| Betasso | 1.5 | EC | 1.52 | 1.1 | 1.6 | 2.1 | 50 | 75 | 0.7 | 1.1 | 1.4 |
| Betasso | 1.5 | UV | 1.69 | 1.1 | 1.7 | 2.3 | 48 | 75 | 0.6 | 1.0 | 1.3 |
| Betasso | 1.5 | GAC | 1.15 | 0.7 | 1.1 | 1.6 | 45 | 72 | 0.6 | 1.0 | 1.4 |
| Betasso | 1.5 | $O_3$/BF | 1.10 | 0.4 | 0.9 | 1.3 | 31 | 69 | 0.3 | 0.8 | 1.1 |
| NPNV | 1.5 | Raw | 5.52 | 3.1 | 5.7 | 7.8 | 41 | 74 | 0.6 | 1.0 | 1.4 |
| NPNV | 1.5 | EC | 3.56 | 1.5 | 2.9 | 4.5 | 34 | 64 | 0.4 | 0.8 | 1.3 |
| NPNV | 1.5 | UV | 3.53 | 1.2 | 2.6 | 4.0 | 29 | 65 | 0.3 | 0.7 | 1.1 |
| NPNV | 1.5 | GAC | 2.59 | 0.8 | 1.9 | 2.8 | 27 | 68 | 0.3 | 0.7 | 1.1 |
| NPNV | 1.5 | O3 | 2.84 | 1.2 | 2.7 | 4.1 | 28 | 65 | 0.4 | 0.9 | 1.5 |
| NPNV | 1.5 | $O_3$/BF | 2.11 | 0.7 | 1.6 | 2.4 | 29 | 65 | 0.3 | 0.7 | 1.1 |
| Winchester | 1.5 | EC | 1.94 | 0.8 | 1.6 | 2.1 | 35 | 73 | 0.4 | 0.8 | 1.1 |
| Winchester | 1.5 | UV | 1.94 | 0.8 | 1.5 | 2.2 | 35 | 68 | 0.4 | 0.8 | 1.1 |
| Winchester | 1.5 | O3 | 1.71 | 0.7 | 1.4 | 2.0 | 33 | 67 | 0.4 | 0.8 | 1.2 |
| Winchester | 1.5 | $O_3$/BF | 1.42 | 0.4 | 0.9 | 1.6 | 25 | 59 | 0.3 | 0.6 | 1.1 |
| Betasso | 2.5 | Raw | 2.86 | 1.9 | 3.6 | 5.1 | 38 | 71 | 0.7 | 1.3 | 1.8 |
| Betasso | 2.5 | EC | 1.52 | 1.3 | 1.9 | 2.7 | 46 | 71 | 0.8 | 1.3 | 1.8 |
| Betasso | 2.5 | UV | 1.69 | 1.2 | 1.9 | 2.9 | 42 | 66 | 0.7 | 1.1 | 1.7 |
| Betasso | 2.5 | GAC | 1.15 | 0.8 | 1.3 | 1.8 | 41 | 70 | 0.7 | 1.1 | 1.6 |
| Betasso | 2.5 | $O_3$/BF | 1.10 | 0.5 | 1.0 | 1.6 | 32 | 61 | 0.5 | 0.9 | 1.4 |
| NPNV | 2.5 | Raw | 5.52 | 3.5 | 6.6 | 9.5 | 37 | 70 | 0.6 | 1.2 | 1.7 |
| NPNV | 2.5 | EC | 3.56 | 1.5 | 3.4 | 5.3 | 29 | 64 | 0.4 | 0.9 | 1.5 |
| NPNV | 2.5 | GAC | 2.59 | 0.9 | 2.2 | 3.3 | 27 | 65 | 0.4 | 0.8 | 1.3 |
| | | | | | | Avg. | 36 | 68 | | | |
| | | | | | | $\sigma$ | 7 | 4 | | | |

consumed in the first hour of decay, and between 61 and 75 percent (avg. = 68 percent) was consumed in the first 24 hours.

### *Effect of Dose*

When the same water was chlorinated at different doses, the chlorine demands and specific chlorine demands ($CD_t$/TOC) at each point in time increased with increasing dose (Table 3.9). However, the percent chlorine consumed at 1 hour and 24 hours with respect to the 120-hour demand were similar for the different doses of the same water. Therefore, the kinetic data at different doses were compared by normalizing the demand by the endpoint. The chlorine demand data were plotted as a fraction of the 120-hour chlorine demand ($CD/CD_{120}$) or as a fraction of the 24-hour chlorine demand ($CD/CD_{24}$) for the different doses of each water. An example of 120-hour normalization is shown for two different doses of Betasso $O_3$/BF water in Figure 3.16, and an example of 24-hour normalization is shown for three different doses of NPNV, EC water in Figure 3.17.

As exemplified by Figures 3.16 and 3.17, the chlorine dose did not affect the chlorine decay kinetics when expressed as a fraction of the long-term demand for any treatment level. In other words, the relative chlorine decay kinetics were the same at either dose for both the short-term and long-term reaction periods. This was true for all treated waters for both Betasso and NPNV source waters, and Koechling (1998) found this same behavior for several raw waters. Therefore, investigation of dose was dropped from the experiments on the treated waters for Winchester.

### *Effect of Treatment*

Enhanced coagulation slightly decreased the specific chlorine demand for each time point at both doses for NPNV, but slightly raised or did not affect the specific chlorine demands for Betasso water (Table 3.9). UV treatment did not affect the TOC of any of the waters, and it also did not affect the chlorine demand of EC water from either Betasso or Winchester waters. However in NPNV water, UV lowered the chlorine demand and specific chlorine demand at both doses. GAC treatment slightly decreased the specific chlorine demand for both waters tested. Ozonation increased the specific chlorine demand of EC water for all three experiments. In NPNV water, at both doses, the increase was substantial at all points in time. In Winchester, the increase was only slight and was not visually observable until 120 hours. The increase in specific chlorine demand due to ozonation suggests that ozone reacts with organic matter to create chlorine reactive sites, and these reactive sites are more prevalent than the amount of TOC completely oxidized by the ozone. Biofiltration lowered the specific chlorine demand of the ozonated water below that of EC water for all three waters and all doses. This means that biofiltration removes some chlorine-reactive compounds originally in the EC water, as well as those created by ozonation.

Figures 3.18 through 3.21 show the effect of treatment on normalized chlorine demand kinetics ($CD/CD_{120}$) for Betasso 1.5:1 $Cl_2$:TOC dose ratio, Betasso 2.5:1 $Cl_2$:TOC dose ratio, NPNV 1.5:1 $Cl_2$:TOC dose ratio, and Winchester 1.5:1 $Cl_2$:TOC dose ratio. Unlike the chlorine kinetics for different chlorine doses, normalizing the curve this way did not consistently unify the curves for differently treated waters. For Betasso water at both chlorine doses (Figures 3.18 and 3.19), EC water displayed faster relative kinetics than the raw water. For NPNV water, the opposite was true. For Betasso, GAC treatment lowered the relative chlorine consumption rate. For both Betasso and Winchester waters, $O_3$/BF water decayed the more slowly relative to the long-term demand. For NPNV none of the advanced treatment processes had an effect on the normalized demand curve.

Figures 3.22 through 3.25 show the effect of treatment on chlorine decay normalized by the 1-hour chlorine residual ($C/C_1$) for Betasso 1.5:1 $Cl_2$:TOC dose ratio, Betasso 2.5:1 $Cl_2$:TOC dose ratio, NPNV 1.5:1 $Cl_2$:TOC dose ratio, and Winchester 1.5:1 $Cl_2$:TOC dose ratio. When presented in this manner, most of the treated waters fall on the same curve, suggesting that the change in chlorine residual due to different treatment processes occurs predominantly in the first hour of decay. In other words, treatment processes affect the fast-reacting compounds. However, there were some exceptions to this trend. For Betasso and Winchester waters, the $O_3$/BF water continued to be more stable than the other waters in the long-term reaction period. For NPNV, Raw and $O_3$ waters were much more reactive than the other waters over the entire decay curve.

### *Effect of Ductile Iron Pipe Sediments*

Figure 3.26 shows the effect of ductile iron pipe sediment on chlorine consumption normalized by the three-hour demand ($CD/CD_3$). The amount of ground sediment (0.5 and 1.0 g/L) did not greatly affect the normalized demand curve. The waters containing the ground sediment decayed very rapidly in the first three hours, consuming 74 percent and 90 percent of their 24-hour demands after three hours for 0.5 and 1.0 g/L of sediment, respectively. The water without sediment exerted its chlorine demand more slowly, consuming only 22 percent of $CD_{24}$ at 3-hours.

## Chlorine Decay Modeling

### *Practical Chlorine Decay Models*

Regression analysis was shown to be successful in developing relationships for predicting chlorine demands after 1, 24, and 120 hours of chlorination. In this section, the term chlorine "demand" is defined as the chlorine consumed at a single point in time.

Based on the data analysis of the effects of dose and treatment on chlorine decay, some practical models predicting chlorine decay have been developed. In this section, the term chlorine "decay" is used to refer to the chlorine residual curve over time.

The models can predict chlorine decay at different doses or after different treatment processes based on a single disinfectant decay study and either the results of a simple demand test (a single measurement at one point in time) at the new conditions or a chlorine demand prediction from relationships developed with regression analysis.

### *Chlorine Demand Relationships*

Koechling (1998) used regression analysis to develop a relationship for chlorine demand as a function of TOC and UVA for raw waters dosed at a 2.0:1 $Cl_2$:TOC dose ratio. A similar approach was used in this study. An additional term, chorine dose, was added to the relationship because these waters included three different $Cl_2$:TOC dose ratios. Equation 3.4 is the general form for the chlorine demand relationship.

**Table 3.10**
**Linear regression of chlorine demands**

| Time (hr) | Coefficients $\chi$ (-) | $\tau$ (-) | $\upsilon$ (cm·mg/L) | Percent contribution $\chi$ ($C_o$) | $\tau$ (TOC) | $\upsilon$ (UVA) | N | $R^2$ |
|---|---|---|---|---|---|---|---|---|
| 1 | 0.109 | 0.072 | 8.73 | 45 | 18 | 37 | 21 | 0.60 |
| 24 | 0.216 | 0.342 | 6.23 | 44 | 43 | 13 | 21 | 0.91 |
| 120 | 0.287 | 0.595 | 9.78 | 41 | 47 | 13 | 18 | 0.95 |

$$CD_t = \chi\ (C_0) + \tau\ (TOC) + \upsilon\ (UVA) \tag{3.4}$$

where $CD_t$ = chlorine demand at time t = 1, 24, or 120 hours (mg/L)
$C_0$ = chlorine dose (mg/L)
TOC = total organic carbon concentration (mg/L)
UVA = ultraviolet absorbance at 254 nanometers ($cm^{-1}$)
$\chi$, $\tau$, $\upsilon$ = regression coefficients, $\chi$,$\tau$ (dimensionless), $\upsilon$ (cm·mg/L)

The results from all the single chlorination experiments in this study for all three source waters, all treatment levels, and all three doses (pH 8.0, 20°C) were used to fit the chlorine demand regressions with a few exceptions.

- None of the raw water experiments (Betasso and NPNV, 1.5:1 and 2:5:1 $Cl_2$:TOC dose ratios) were used in determining the regression coefficients because their high CD values exerted too much leverage on the regression.
- NPNV $O_3$ experiments were not used in the regression because their behavior did not fit with the other waters. As shown in Table 3.9, the chlorine demand to TOC ratio for NPNV $O_3$ water was significantly higher than all other waters over the entire decay period, especially at the higher $Cl_2$:TOC dose ratio. All of the other waters showed a positive correlation of both TOC and UVA with chlorine demand, but the decrease in TOC (12%) and UVA (78%) due to ozonation did not correspond to a decreased chlorine demand for NPNV. Winchester $O_3$ water was only dosed at the base case 1.5:1 $Cl_2$:TOC dose ratio. At this dose it did not exhibit the same outlying chlorine demand behavior as NPNV $O_3$ and was left in the regression data set. The SUVA of the Winchester EC water was 16% higher than the NPNV EC water and ozonation removed ten percent less UVA. Chlorine decay experiments were not performed on Betasso ozonated water except after biofiltration.

Table 3.10 shows the resulting regression coefficients for each chlorine demand time.

Chlorine dose contributes more than TOC and UVA to the 1-hour and 24-hour chlorine demand regressions. However, TOC contributes slightly more than dose to the 120-hour regression. The 1-hour chlorine demand is the most dependent on $UVA_{254}$, which corresponds with raw water results of Koechling (1998) that UV absorbing organics react fast with chlorine.

Figures 3.27 through 3.29 show predicted versus actual values for $CD_1$, $CD_{24}$, and $CD_{120}$ for all of the single-chlorination experiments performed in this study. A line for the 1:1 Regression versus Actual CLMD is shown on Figures 3.27 through 3.29 to serve as a visual representation of where the symbols would ideally fall. The hollow symbols represent the data not used to predict the regression (Raw and NPNV $O_3$ waters) and were not used in calculating the $R^2$ values shown. The 1-hour regression was the least successful at fitting the chlorine demand, while the 24-hour and 120-hour regressions were well fit. It is interesting that the discarded Raw chlorine demands were predicted well at all three times, with a maximum error of 0.1 mg/L for $CD_{24}$ (Betasso Raw 1.5:1 $Cl_2$:TOC). The chlorine demand of NPNV $O_3$ tests were predicted within 0.5 mg/L for 1 hour, and 0.3 mg/L for 24 and 120 hours.

The overall success of these linear regressions indicates that chlorine demands can be predicted for any water, treated or untreated, dosed at a $Cl_2$:TOC dose ratio in the range of 0.75 to 2.5 using Equation 3.4 and the coefficients in Table 3.10.

### *Predicting Chlorine Decay for New Doses*

As discussed previously, at any point in time the fraction of the chlorine demand ($CD_t$) consumed, with respect to the total long-term chlorine demand ($CD_{24}$ or $CD_{120}$), is the same for any dose in the same water. Therefore, all that is needed to predict the decay of chlorine at new doses is a known chlorine decay curve at any dose and the long-term chlorine demand (24 hours or 120 hours) at the new dose. This can either be measured experimentally or predicted via Equation 3.4 and Table 3.10. Equation 3.5 can then be used to predict the chlorine residual at any point in time for the new dose conditions.

$$C_t = C_0 - \left(\frac{CD_t^*}{CD_{120}^*}\right) \cdot CD_{120} \qquad (3.5)$$

where $C_t$ = chlorine residual at time, t, at new chlorine dose (mg/L)
$C_0$ = new chlorine dose (mg/L)
$CD_t^*$ = chlorine demand at time, t, for old chlorine dose (mg/L)
$CD_{120}^*$ = chlorine demand at 120 hours for old chlorine dose (mg/L)
$CD_{120}$ = chlorine demand at 120 hours for new chlorine dose (mg/L)

Equation 3.5 was used to predict the chlorine residual curve at 2.5:1 $Cl_2$:TOC dose ratio from the known decay curve at 1.5:1 $Cl_2$:TOC dose ratio for for Betasso, GAC water (pH 8.0, 20°C). Predictions were made using the measured $CD_{120}$ at the $Cl_2$:TOC dose ratio and the predicted $CD_{120}$ using Equation 3.4 and the values in Table 3.10. The model results and the experimental data are compared in Figure 3.30. Similarly accurate results were found for all of the other waters.

### *Predicting Chlorine Decay After Treatment*

As discussed previously, the fraction of the chlorine residual ($C_t$) with respect to the 1-hour chlorine residual ($C_1$) is the same for treatment levels of raw, EC, GAC, or $O_3$/BF in the same water at any point in time. Therefore, all that is needed to predict the decay of chlorine after treatment is a known chlorine decay curve at any treatment level and the 1-hour chlorine residual ($C_1$) at the new treatment level. This can either be measured experimentally or predicted via Equation 3.6.

$$C_1 = C_0 - CD_1 \tag{3.6}$$

where $C_1$ = chlorine residual in water at 1 hour after new treatment (mg/L)
$C_0$ = initial chlorine concentration (mg/L)
$CD_1$ = 1-hour chlorine demand predicted via Equation 3.4, Table 3.10.

Equation 3.7 can then be used to predict the chlorine residual at any point in time for the new treatment conditions.

$$C_t = C_1 - \left(\frac{C_t^*}{C_1^*}\right) \tag{3.7}$$

where $C_t$ = chlorine residual at time t after new treatment (mg/L)
$C_1$ = chlorine residual at 1 hour after new treatment (mg/L)
$C_t^*$ = chlorine residual at time t before new treatment (mg/L)
$C_1^*$ = chlorine residual at 1 hour before new treatment (mg/L)

Equation 3.7 was used to predict the chlorine residual curve after GAC and $O_3$/BF treatment from the known EC water chlorine decay curves for each source water using the known 1-hour chlorine residuals. Figures 3.31, 3.32, and 3.33 show the results of the model for Betasso, NPNV, and Winchester waters, respectively (1.5:1 $Cl_2$:TOC dose, pH 8.0, 20°C). The chlorine decay for all of the treated waters were predicted very well, except for Betasso $O_3$/BF, which was more stable than the model predicted.

Figures 3.34, 3.35, and 3.36 show the same scenarios, except that the models use predicted values for $C_1$ (Equations 3.4 and Equation 3.6, Table 3.10). The 1-hour chlorine residuals for the NPNV and Winchester treated waters were predicted very well, and the chlorine decay curves were also accurate. For Betasso $O_3$/BF water, the $C_1$ value was predicted within 0.01 mg/L of the actual value. However, as before, the model overestimated the chlorine decay rate. For Betasso GAC water, the $C_1$ value was over predicted, throwing off the decay curve.

***Numerical Chlorine Decay Models***

As discussed in Chapter 2, Koechling (1998) developed a Michaelis-Menton chlorine decay model based on raw waters dosed at a 2.0:1 $Cl_2$:TOC dose ratio and Isabel et al. (2000) amended the model to account for a range of dose ratios (outlined in Table 2.1). Figure 3.37 shows the results of this model for Betasso EC water (2.5:1 $Cl_2$:TOC, pH 8.0, 20°C), using both the Koechling/Isabel predicted values for the rate constants $\alpha_1$ and $k_2$, and the best-fit values for these constants. The prediction-based model under predicted the chlorine decay for the entire reaction period. The long-term chlorine decay (after three hours reaction time) was well described by the best-fit model, but the first three hours of decay were not. The model exhibited a low reaction rate in the short-term period that is not descriptive of actual chlorine decay, which reacts very quickly in the first few hours of decay.

Koechling (1998) also developed a temperature correction for the rate constant, $k_2$, based on an Arrhenius temperature relationship (Equation 3.8) with $\theta = 1.10$.

$$k_2(T) = k_2(20°C) \cdot \theta^{(T-20)} \tag{3.8}$$

where $k_2(T)$ = rate constant, $k_2$, at temperature, T (mg/L)
$k_2(20°C)$ = rate constant, $k_2$, at 20°C (mg/L)
$\theta$ = Arrhenius coefficient(-)
T = temperature (°C)

Figure 3.38 shows the Koechling (1998) model predictions for chlorine decay at temperatures of 4 and 30°C based on the best-fit rate constants for 20°C (Betasso, EC, 2.5:1 $Cl_2$:TOC, pH 8.0). The resulting models predicted the change in chlorine decay due to temperature very well, especially for the 30°C test.

**Chloramine Decay Primary Results**

The primary results of this study are presented in terms of chloramine residual concentration versus time. In this section, the effects of the different variables on these results are discussed. For many of the scenarios, additional data analysis is provided in the next section. All primary data for chloramines are reported in Appendix B. Figure 3.39 shows chloramine residual as a function of time. Under base case conditions (pH 8.0, and 20°C) chloramine is more stable than chlorine, and the decay rate is much lower. While some initial demand is noted, it is not as significant as that found with chlorine. Chloramine decay measurements were extended to a period of three weeks. To facilitate the chloramine decay kinetics presentation over this period, semi-logarithmic plots are presented.

***Effect of Dose***

Figure 3.40 presents the same data as in Figure 3.39, showing the effect of dose. The chloramine dose for each water was selected such that the 24-hour chloramines residual was either 2 or 4 mg($Cl_2$)/L for all treatment conditions. This plot is representative of the results found for all test conditions. In the case of Betasso water, chloramine decay studies were conducted targeting 2, 3, and 4 mg/L residuals at 24 hours. Further discussion of the effect of dose on the chloramines decay is in the data analysis section.

***Effect of pH***

The pH of chloraminated waters is normally kept above 8.0 to maintain the dominance monochloramine over dichloramine. To assess the effect of higher pH on chloramine decay, pH studies in which the pH was raised to 10.5 were conducted with four different waters; EC Betasso, GAC Betasso, EC NPNV, and EC Winchester. For each study, a non-systematic effect of pH was found. The pH study conducted on EC NPNV (Figure 3.41) shows that water with a pH of 8.0

decayed faster than the water with a pH of 10.5 over the long term. This behavior was expected based on the chloramine autodecomposition models proposed by Vikesland et al. (2001). Contrary to the findings from EC NPNV water, the GAC Betasso water with a pH value of 10.5 decayed faster after 8 hours than the water with a lower pH value of 8.0 (Figure 3.42). These findings are not what would be predicted based on the chloramine autodecomposition model. Figure 3.43 shows the effect of pH on chloramine decay for EC Winchester water. In this decay study, little difference in the decay kinetics of waters at pH 8.0 or 10.5 were found. It can be seen in the Appendix B (raw data tables) that the pH studies conducted on EC Betasso water also showed little difference in the chloramine decay kinetics between the waters with pH 8.0 and 10.5. The only trend observed is that the highest TOC water with the largest chloramine demand, EC NPNV, yielded the expected behavior. However, it was expected that this water would deviate from the autodecomposition model as the NOM reduction of the chloramine would be the greatest.

***Effect of Temperature***

Figure 3.44 shows the effect of temperature on chloramine decay kinetics for $O_3$/BF Winchester water. After 8 hours, a systematic difference in chloramine decay becomes apparent. As expected, the high temperature studies resulted in faster chloramine decay, and the low temperature studies yielded the most stable residuals. Six temperature studies were conducted on different waters; EC Betasso, $O_3$/BF Betasso, EC NPNV, $O_3$/BF NPNV, and EC Winchester. For all six experiments, the effect of temperature was similar at the various treatment levels.

***Effect of Treatment***

The effect of treatment was evaluated for all three source waters at all treatment levels; EC, GAC, UV, and $O_3$/BF. In addition, the chloramine decay behavior of ozone treated water for both NPNV water and Winchester water was studied. The effect of treatment on Winchester water at a 2 mg/L target chloramine residual shows that the ozonated water was the most reactive over the long term (Figure 3.45). Other treatment levels for this water showed no difference in chloramine decay behavior. Similar to behavior of the Winchester water, the ozone treated NPNV water (Figure 3.46) appeared to be most reactive, as well as the raw water, while UV irradiation made the water more stable with respect to chloramine reactions.

***Effect of Blending Chloraminated Waters***

To evaluate the effect of mixing the same water but with different ages and chloramine residuals on chloramine decay, two GAC Betasso waters that had been reacted with chloramines for either 5 days or 2 hours were blended at a 1:1 volume ratio (Figure 3.47). After blending the 5-day "old water" with the 2-hour "new water," the chloramine decay behavior was monitored for 3 weeks. Because of the small difference in the chloramine residuals at 2 hours (2.78 mg/L) and 5 days (2.70 mg/L), the blended water behaves similar to that of the unblended GAC water, which is stable out to three weeks.

To assess the effect of blending the different waters with very different chloramine residuals, two different waters that targeted two different 24-hour chloramine residuals were blended. In this scenario, EC water from the 63rd Street WTP (City of Boulder) was dosed with chloramine to target 2 mg/L at 24 hours. Simultaneously, EC NPNV water was dosed with chloramine to target 4 mg/L at 24 hours. After the waters had reacted for 2 hours, they were blended and the chloramine behavior was again tracked for 3 weeks as shown in Figure 3.48. The results showed that the blended water behaved as the average of the chloramine residual concentrations of the two different waters. This average calculation is shown as a dashed line in the figure.

The effect of blending waters with different chloramine demands on the chloramine decay was assessed by blending two waters with different TOC concentrations and dosed to yield the same 24-hour chloramine residual. Both EC 63rd Street and EC NPNV water were reacted with chloramine for 2 hours before blending the two waters together and tracking the chloramine behavior for 3 weeks. The results are shown in Figure 3.49. Similar to the previous blending studies discussed, this study shows that the effect of demand on blended waters behaves as a weighted average of the different demands used in this experiment, shown as the dashed line.

### *Effect of Rechloramination*

The effect of rechloramination on chloramine decay was studied by boosting the chloramine concentration of chloraminated water that had reacted for 7 days with a solution of preformed chloramine. The EC NPNV water was dosed at 3 mg/L and at 4 mg/L. After 5 days, 1 mg/L of chloramine was added to the solution that was originally dosed at 3 mg/L, yielding the same total chloramine addition for both solutions. The results are shown in Figure 3.50. The rechloraminated water displayed nearly the same decay for 10 additional days as the solution dose that was originally at 4 mg/L. Similar chloramine decay kinetics were found with the rechloramination of the EC Winchester water at a total dose of 3.8 mg/L.

## Chloramine Decay Data Analysis

### *Effect of Dose*

Figure 3.51 illustrates the normalization of the chloramine residual by the initial dose for the $O_3$/BF Betasso water at three initial doses. The relative decay kinetics were found to be independent of dose. Similar dose studies were conducted for all waters at nearly all treatment levels yielding 17 experiments in which two or three doses were used. This approach of normalizing by dose yields very good agreement for nine of the 17 waters. Of the remaining eight experiments, four showed similar behavior with the exception of 20-day chloramine residuals, where small variations were found.

### *Effect of Treatment*

To analyze the effect of treatment, the chloramine residuals were normalized by the dose. The results for the Betasso water at a target residual of 4 mg/L (Figure 3.52) show that raw and UV treated waters are the most reactive waters for the low dose over the long term, whereas the GAC and $O_3$/BF treated waters are the most stable over the long term. Figure 3.53 shows the results for the NPNV water at a target residual of 4 mg/L. In this case, UV irradiation of the water makes it the most stable of the treatment levels. Similar to the results found with Betasso water, the raw water is the most reactive. For this water, the EC water was also treated by ozone alone yielding a reactive water over the 3-week test period. By biofiltering the ozonated water, the chloramine decay behavior stabilized. The results for the Winchester water are shown in

**Table 3.11**
**Bench-scale chloramine demand results**

| Source water | Target residual (mg/L) | Treatment level | TOC (mg/L) | Initial dose (mg/L) | Chloramine demand $CLMD_t$ (mg/L) | | | Specific demand $CLMD_t$:TOC (mg/mg) | | |
|---|---|---|---|---|---|---|---|---|---|---|
| | | | | | 24-hr | 5-day | 20-day | 24-hr | 5-day | 20-day |
| Betasso | 2 | Raw | 2.86 | 2.9 | 0.51 | 0.68 | 1.41 | 0.18 | 0.24 | 0.49 |
| Betasso | 2 | EC | 1.52 | 2.5 | 0.36 | 0.31 | 0.73 | 0.24 | 0.20 | 0.48 |
| Betasso | 2 | GAC | 1.15 | 2.2 | 0.24 | 0.27 | 0.57 | 0.21 | 0.23 | 0.50 |
| Betasso | 2 | UV | 1.69 | 2.5 | 0.51 | 0.54 | 0.92 | 0.28 | 0.30 | 0.51 |
| Betasso | 2 | $O_3$/BF | 1.10 | 2.2 | 0.07 | 0.14 | 0.84 | 0.06 | 0.13 | 0.76 |
| Betasso | 3 | Raw | 2.86 | 4.3 | 0.85 | 1.28 | 2.00 | 0.30 | 0.45 | 0.70 |
| Betasso | 3 | EC | 1.52 | 3.5 | 0.38 | 0.93 | 1.41 | 0.25 | 0.61 | 0.93 |
| Betasso | 3 | GAC | 1.15 | 3.3 | 0.14 | 0.23 | 0.54 | 0.12 | 0.20 | 0.47 |
| Betasso | 3 | $O_3$/BF | 1.10 | 3.3 | 0.09 | 0.31 | 0.68 | 0.08 | 0.28 | 0.62 |
| Betasso | 4 | Raw | 2.86 | 5.8 | 1.09 | 1.50 | 2.58 | 0.38 | 0.52 | 0.90 |
| Betasso | 4 | EC | 1.52 | 5.6 | 0.75 | 0.95 | 1.50 | 0.49 | 0.63 | 0.99 |
| Betasso | 4 | GAC | 1.15 | 4.5 | 0.14 | 0.28 | 0.70 | 0.12 | 0.24 | 0.61 |
| Betasso | 4 | UV | 1.69 | 5.6 | 0.96 | 1.16 | 1.75 | 0.54 | 0.65 | 0.98 |
| Betasso | 4 | $O_3$/BF | 1.10 | 4.5 | 0.21 | 0.50 | 1.00 | 0.19 | 0.45 | 0.91 |
| NPNV | 2 | Raw | 5.52 | 3.2 | 0.98 | 1.55 | 2.44 | 0.16 | 0.26 | 0.41 |
| NPNV | 2 | EC | 3.56 | 3.0 | 0.18 | 0.65 | 1.35 | 0.05 | 0.19 | 0.39 |
| NPNV | 2 | GAC | 2.59 | 2.9 | 0.52 | 0.76 | 1.55 | 0.20 | 0.30 | 0.61 |
| NPNV | 2 | UV | 3.53 | 3.0 | 0.28 | 0.42 | 1.10 | 0.08 | 0.12 | 0.31 |
| NPNV | 2 | $O_3$/BF | 2.11 | 2.9 | 0.40 | 0.73 | 1.31 | 0.19 | 0.24 | 0.73 |
| NPNV | 2 | $O_3$ | 2.84 | 3.0 | 0.88 | 1.68 | 2.18 | 0.31 | 0.59 | 0.77 |
| NPNV | 4 | Raw | 5.52 | 6.5 | 1.90 | 2.40 | 4.11 | 0.32 | 0.40 | 0.69 |
| NPNV | 4 | EC | 3.56 | 5.3 | 1.07 | 1.55 | 2.52 | 0.31 | 0.45 | 0.73 |
| NPNV | 4 | GAC | 2.59 | 5.0 | 0.08 | 1.25 | 2.40 | 0.03 | 0.49 | 0.94 |
| NPNV | 4 | UV | 3.53 | 5.3 | 1.10 | 0.50 | 1.53 | 0.31 | 0.14 | 0.43 |
| NPNV | 4 | $O_3$/BF | 2.11 | 5.0 | 0.80 | 1.45 | 2.18 | 0.38 | 0.68 | 1.03 |
| NPNV | 4 | $O_3$ | 2.84 | 5.3 | 1.05 | 2.16 | 3.11 | 0.37 | 0.76 | 1.10 |
| Winchester | 2 | EC | 1.94 | 2.5 | 0.24 | 0.55 | 0.77 | 0.12 | 0.28 | 0.40 |
| Winchester | 2 | GAC | 1.30 | 2.5 | 0.33 | 0.47 | 0.45 | 0.25 | 0.36 | 0.35 |
| Winchester | 2 | UV | 1.94 | 2.5 | 0.40 | 0.54 | 0.87 | 0.48 | 0.59 | 0.81 |
| Winchester | 2 | $O_3$/BF | 1.42 | 2.5 | 0.30 | 0.55 | 0.80 | 0.21 | 0.39 | 0.56 |
| Winchester | 2 | $O_3$ | 1.71 | 2.5 | 0.46 | 0.70 | 1.12 | 0.27 | 0.41 | 0.66 |
| Winchester | 4 | EC | 1.94 | 4.8 | 0.62 | 0.92 | 1.44 | 0.32 | 0.47 | 0.74 |
| Winchester | 4 | GAC | 1.30 | 4.8 | 0.51 | 0.82 | 0.96 | 0.39 | 0.63 | 0.73 |
| Winchester | 4 | UV | 1.94 | 4.8 | 0.94 | 1.15 | 1.57 | 0.48 | 0.59 | 0.81 |
| Winchester | 4 | $O_3$/BF | 1.42 | 4.8 | 0.67 | 0.96 | 1.17 | 0.47 | 0.68 | 0.82 |
| Winchester | 4 | $O_3$ | 1.71 | 4.8 | 0.65 | 0.85 | 1.61 | 0.38 | 0.50 | 0.95 |

**Table 3.12**
**Linear regression of chloramine demands**

| Time (days) | Coefficients | | | Percent contribution | | | N | $R^2$ |
|---|---|---|---|---|---|---|---|---|
| | $\chi$ (-) | $\tau$ (-) | $\upsilon$ (cm·mg/L) | $\chi$ ($C_o$) | $\tau$ (TOC) | $\upsilon$ (UVA) | | |
| 1 | 0.082 | 1.097 | 1.097 | 58 | 52 | -10 | 35 | 0.68 |
| 5 | 0.182 | 0.319 | 0.495 | 45 | 62 | - 7 | 35 | 0.61 |
| 20 | 0.125 | 0.194 | 1.467 | 49 | 50 | 1 | 35 | 0.77 |

Figure 3.54. Like NPNV results, the ozonated water is the most reactive, and biofiltration of that water yielded more stable chloramine behavior. Treatment beyond EC of this water by UV and GAC yielded little effect on the relative decay kinetics.

Attempts to normalize further the chloramine data by methods used for the chlorine data were not successful. There was little chloramine demand between 15 min and 1 hour. Thus there was no benefit to normalizing the data by the 1-hour residual. The chloramine demand data are shown in Table 3.11. No specific trend could be found for the effect of treatment on the chloramine reactivity (chloramine demand/TOC). This lack of relationships impaired the development of chloramine decay models.

## Chloramine Demand Modeling

As with chlorine, regression analysis was shown to be successful in developing relationships for predicting chloramine demands. Equation 3.9 is the general form for the chloramine demand relationship.

$$CLMD_t = \chi\,(C_0) + \tau\,(TOC) + \upsilon\,(UVA) \qquad (3.9)$$

where $CLMD_t$ = chlorine demand at time, t = 1, 5, or 20 days (mg/L)
$C_0$ = chloramine dose (mg/L)
TOC = total organic carbon concentration (mg/L)
UVA = ultraviolet absorbance at 254 nanometers ($cm^{-1}$)
$\chi$, $\tau$, $\upsilon$ = regression coefficients, $\chi$,$\tau$ (dimensionless), $\upsilon$ (cm·mg/L)

The chloramine demands of 24 hours, 5 days, and 20 days from the single chloramination experiments tested in this study for all three source waters, all treatment levels, and all three doses (pH 8.0, 20°C) were regressed in a similar way. The NPNV raw water experiments were not used in determining the regression coefficients because their high CLMD values exerted too much leverage on the regression. $O_3$ Only experiments were found to be regression outliers and not used in the regression. Table 3.12 shows the resulting regression coefficients for each chloramine demand time.

Chloramine dose and TOC contribute heavily to all three chloramine demand regressions, in almost equal proportions to one another for each time point. The 1-day and 5-day chlorine demands exhibit a slight inverse dependence on $UVA_{254}$, but after 20 days UVA has almost no influence on the regression for chloramine demand. This suggests that single-bonded carbon aliphatic organics react slightly faster with chloramines than double-bonded or aromatic organics.

Figures 3.55 through 3.57 show predicted versus actual values for $CLMD_1$, $CLMD_5$, and $CLMD_{20}$ for all of the single-chloramination experiments performed in this study. A line for the 1:1 Regression versus Actual CLMD is shown on Figures 3.55 through 3.57 to serve as a visual representation of where the symbols would ideally fall. The hollow symbols represent the data not used to predict the regression (Raw and NPNV $O_3$ waters) and were not used in calculating the $R^2$ values shown for the same reasons presented previously when discussing the free chlorine regressions. The 1-day regression was the least successful at fitting the chloramine demand and the 20-day regression was the best. As noted previously, further development of chloramine decay models was impaired because no specific trend could be found for the effect of treatment on the chloramine reactivity (chloramine demand/TOC).

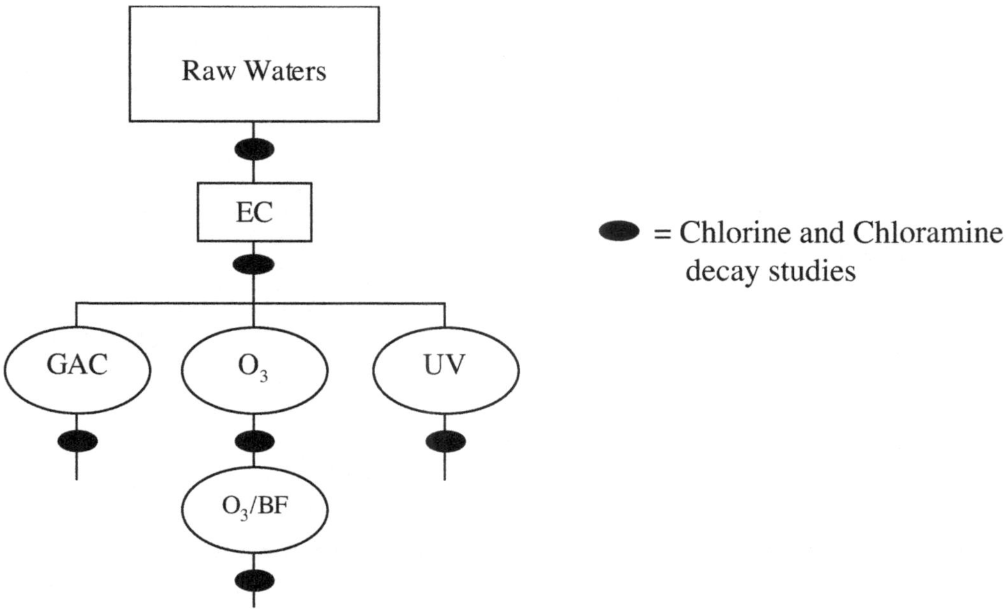


**Figure 3.1 Chlorine and chloramine decay study and treatment flow diagram**

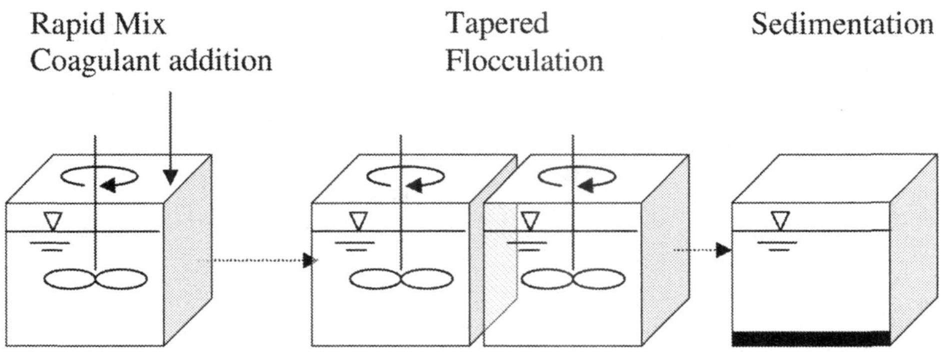


**Figure 3.2 Jar test and EC mixing schematic**

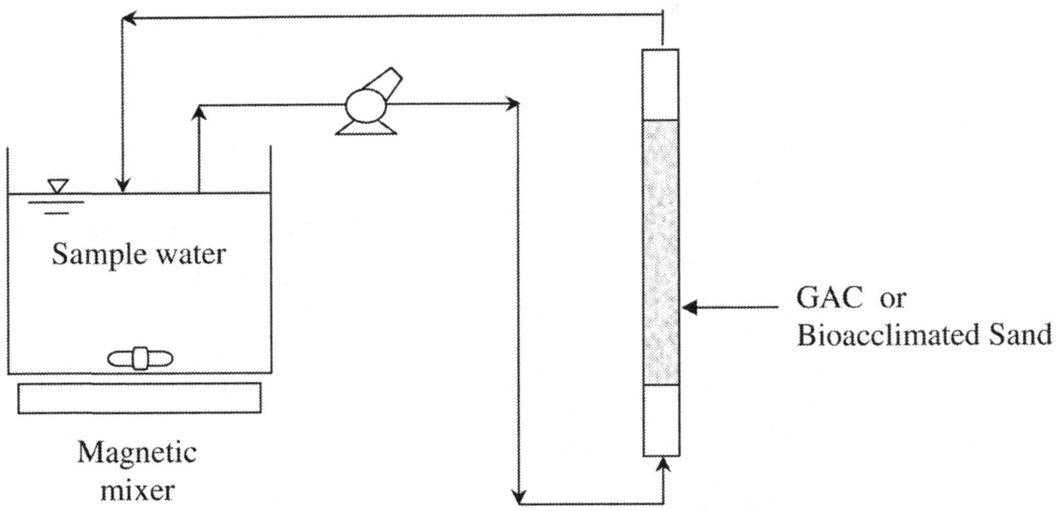


**Figure 3.3 Schematic of recirculating GAC and biofiltration system**

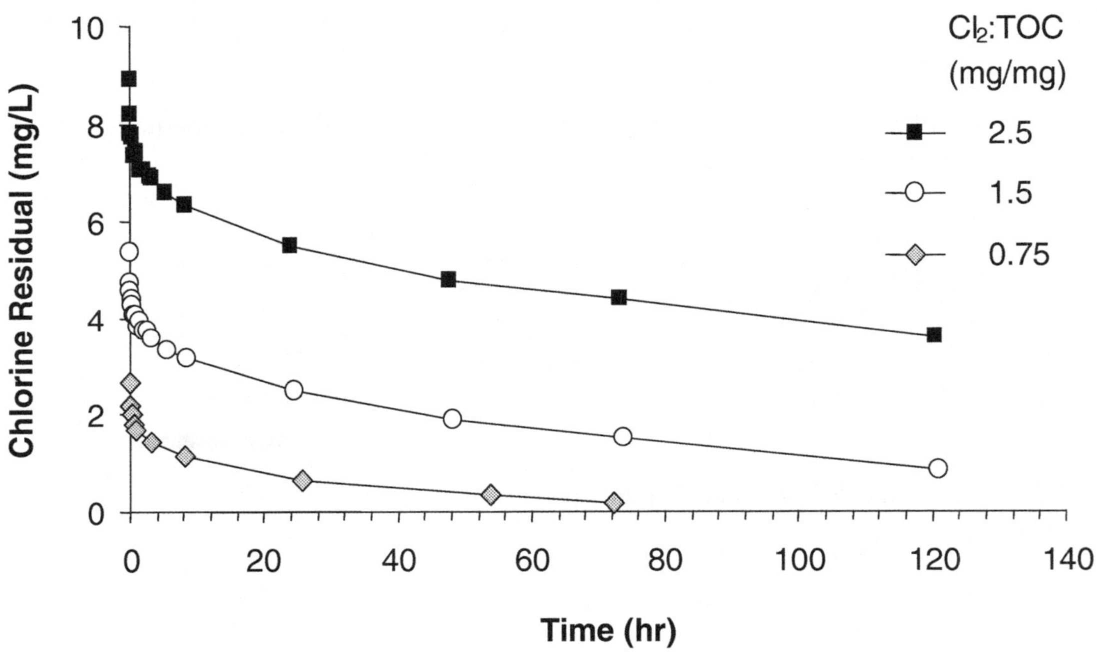


**Figure 3.4 Effect of dose on chlorine decay: arithmetic time scale. NPNV, EC, TOC = 3.56 mg/L, pH 8.0, 20°C.**

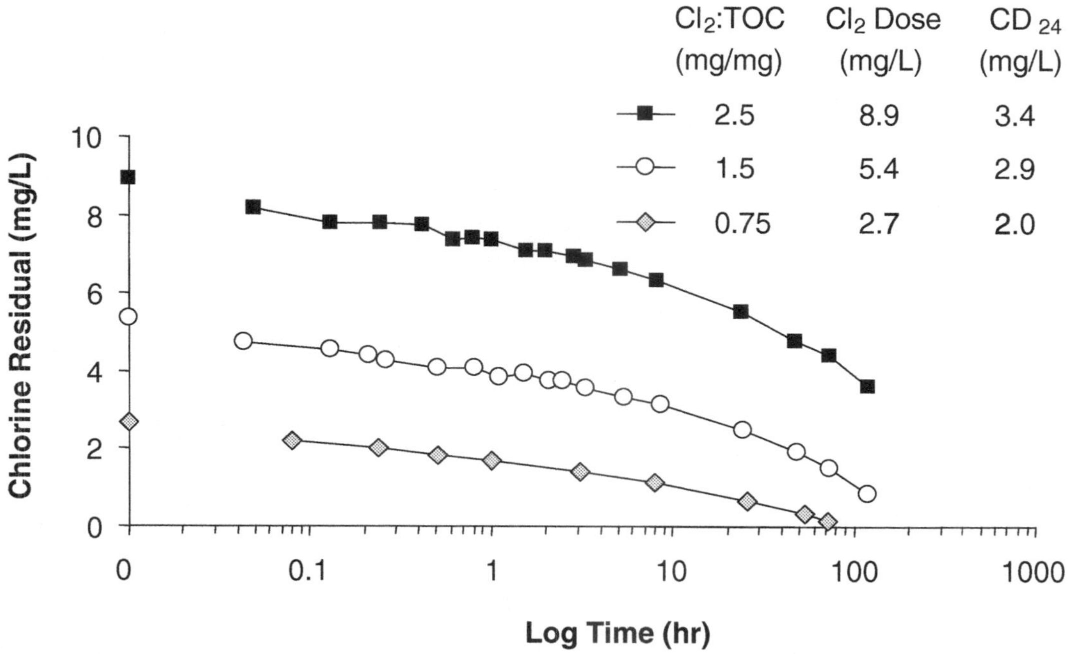


**Figure 3.5 Effect of dose on chlorine decay: logarithmic time scale. NPNV, EC, TOC = 3.56 mg/L, pH 8.0, 20°C.**

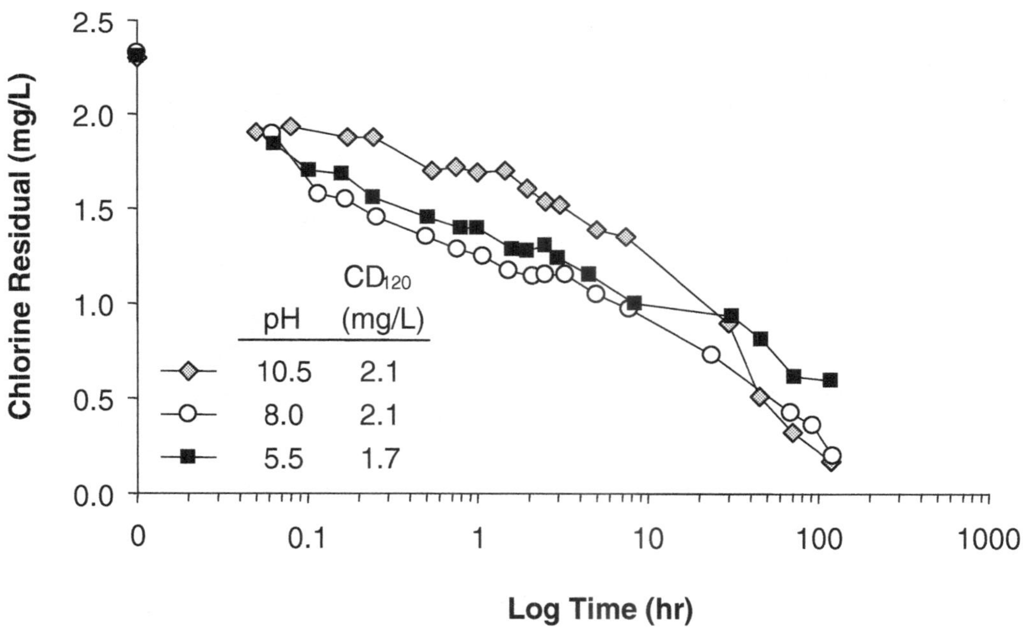


**Figure 3.6 Effect of pH on chlorine decay. Betasso, EC, TOC = 1.52 mg/L, 1.5:1 $Cl_2$:TOC dose (mg/mg), 20°C.**

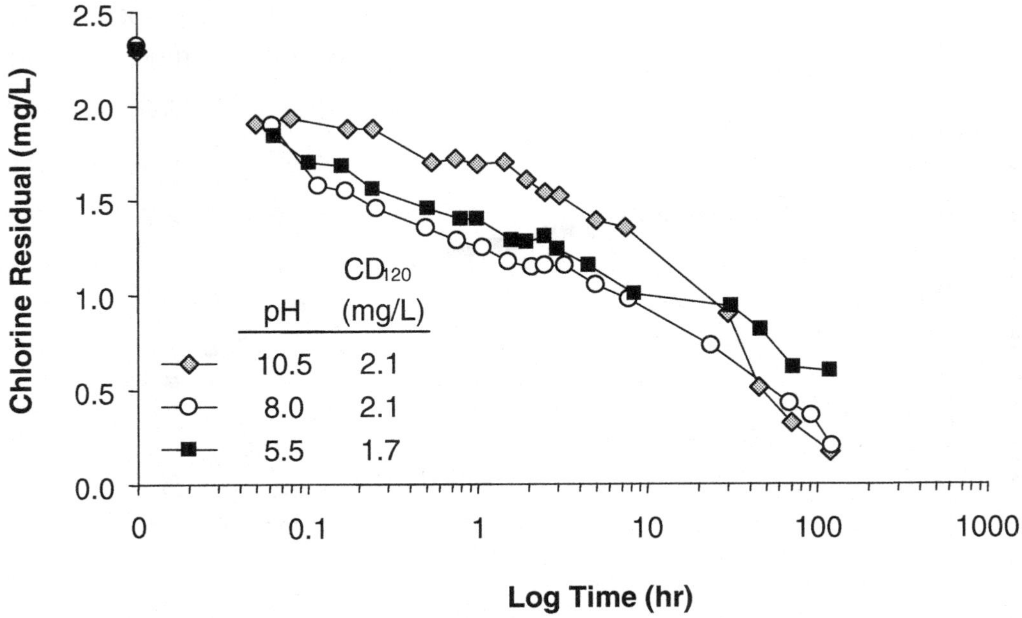


**Figure 3.7 Effect of pH on chlorine decay. NPNV, EC, TOC = 3.56 mg/L, 1.5:1 $Cl_2$:TOC dose (mg/mg), 20°C.**

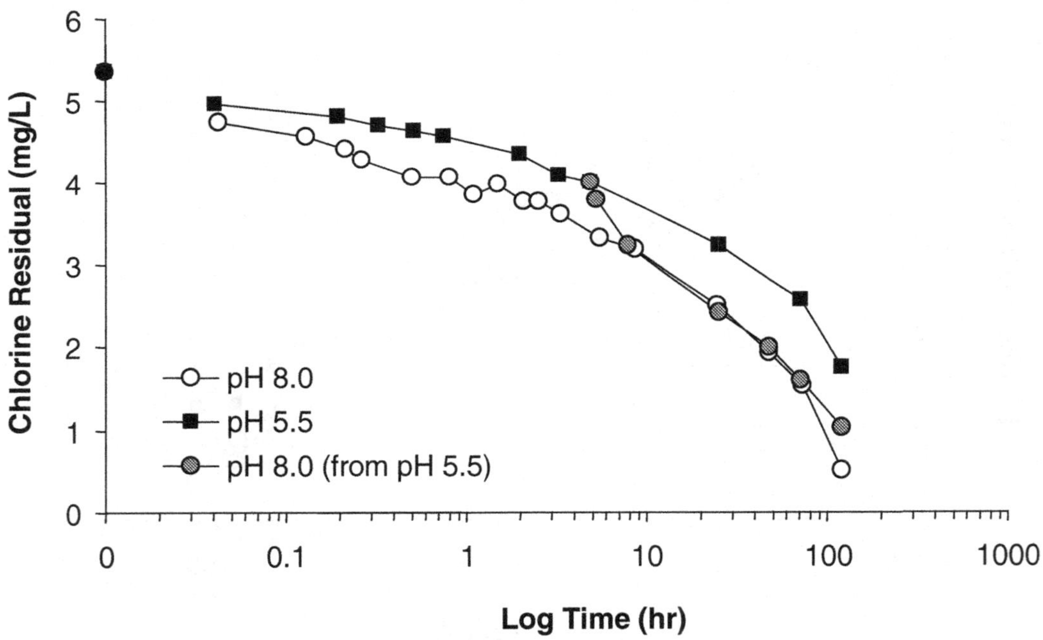


**Figure 3.8 Effect of changing pH on chlorine decay: pH 5.5 from time zero to five hours, pH 8.0 from time five hours to 5 days. NPNV, EC, TOC = 3.56 mg/L, 1.5:1 $Cl_2$:TOC dose (mg/mg), 20°C.**

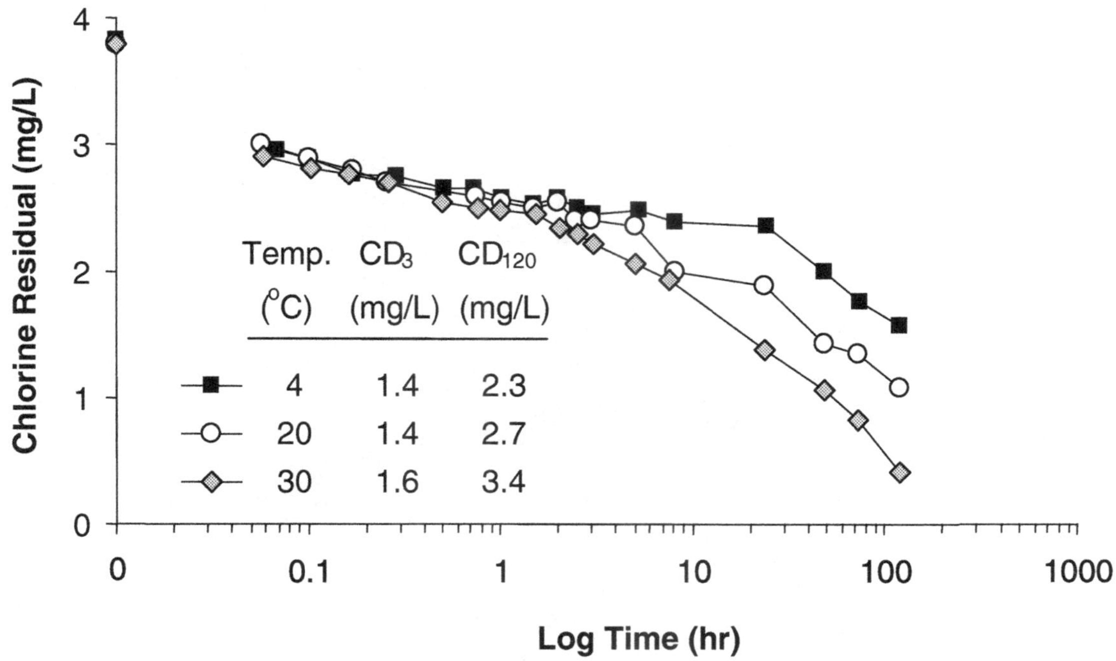


**Figure 3.9 Effect of temperature on chlorine decay. Betasso, EC, TOC = 1.52 mg/L, 2.5:1 $Cl_2$:TOC dose (mg/mg), pH 8.0.**

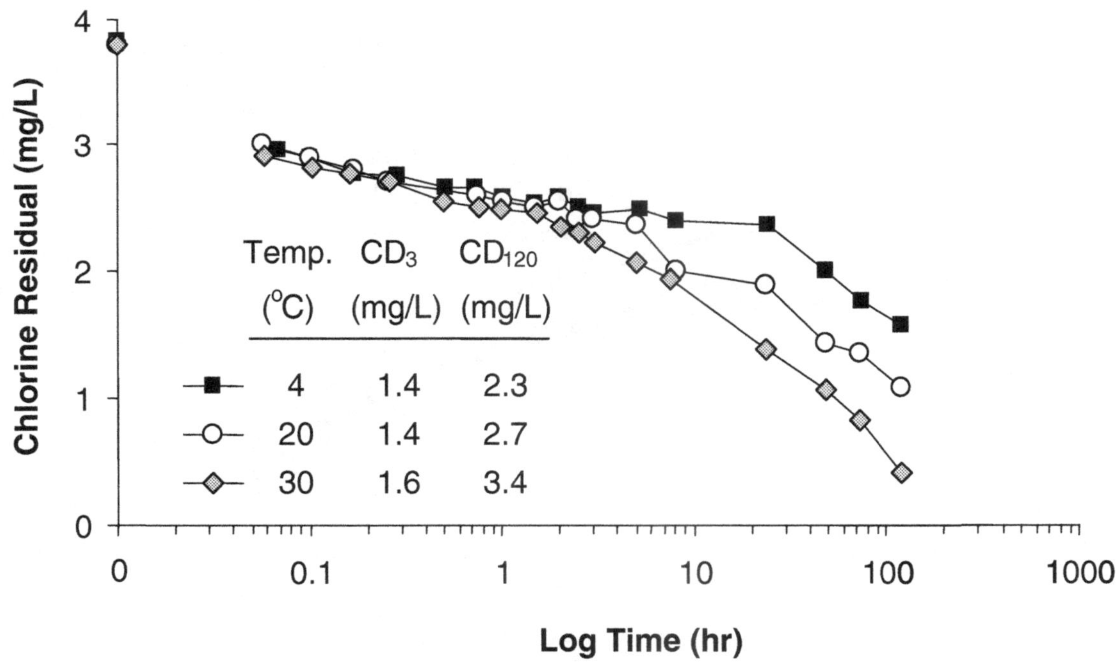


**Figure 3.10 Effect of treatment on chlorine decay. Betasso, 1.5:1 $Cl_2$:TOC dose (mg/mg), pH 8.0, 20°C.**

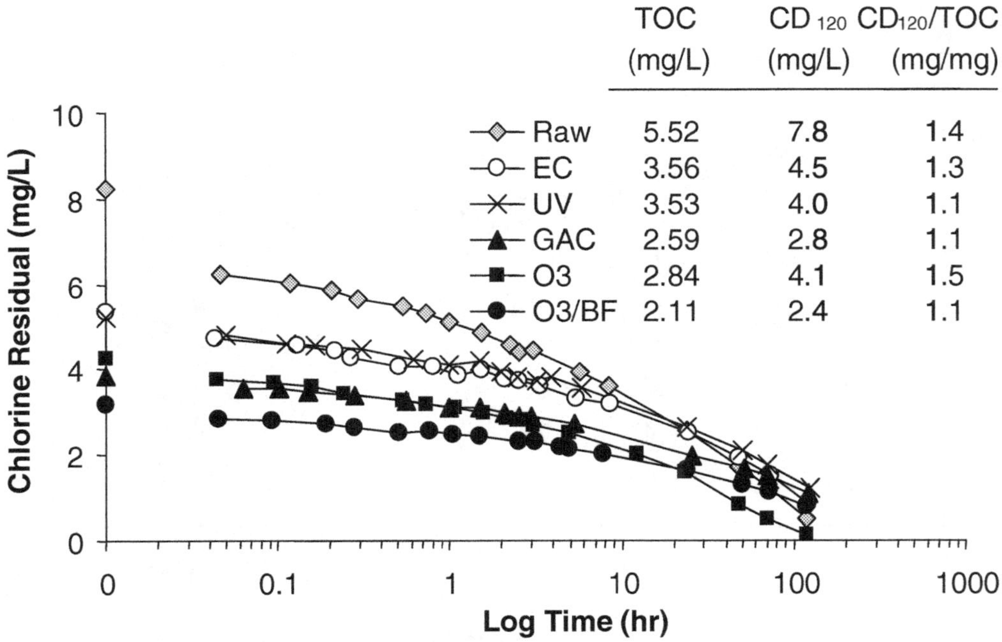


**Figure 3.11 Effect of treatment on chlorine decay. NPNV, 1.5:1 $Cl_2$:TOC dose (mg/mg), pH 8.0, 20°C.**

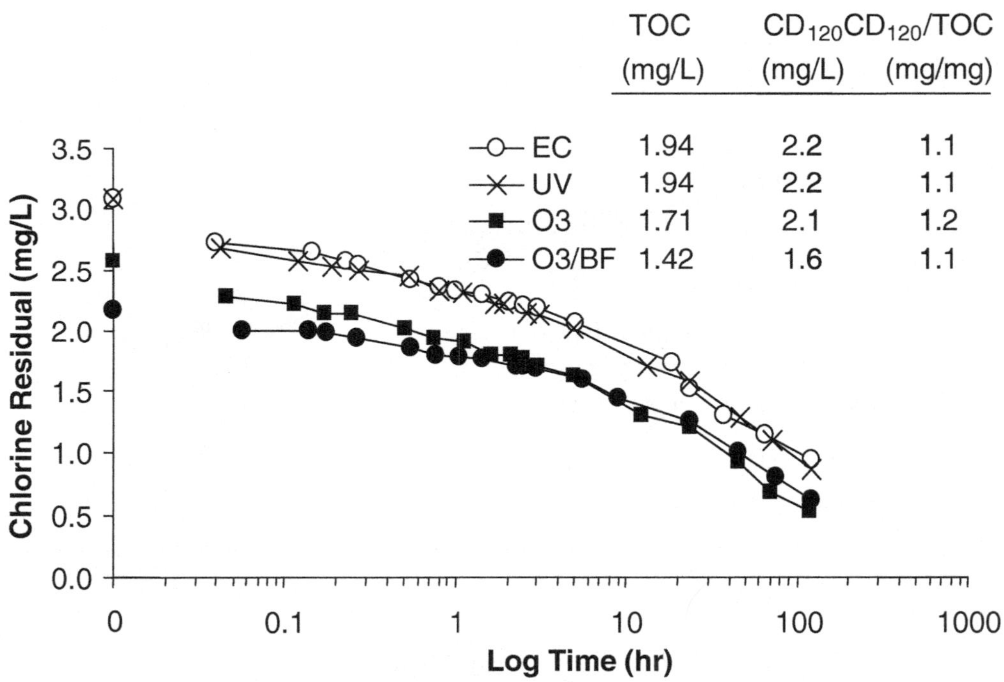


**Figure 3.12 Effect of treatment on chlorine decay. Winchester, 1.5:1 $Cl_2$:TOC dose (mg/mg), pH 8.0, 20°C.**

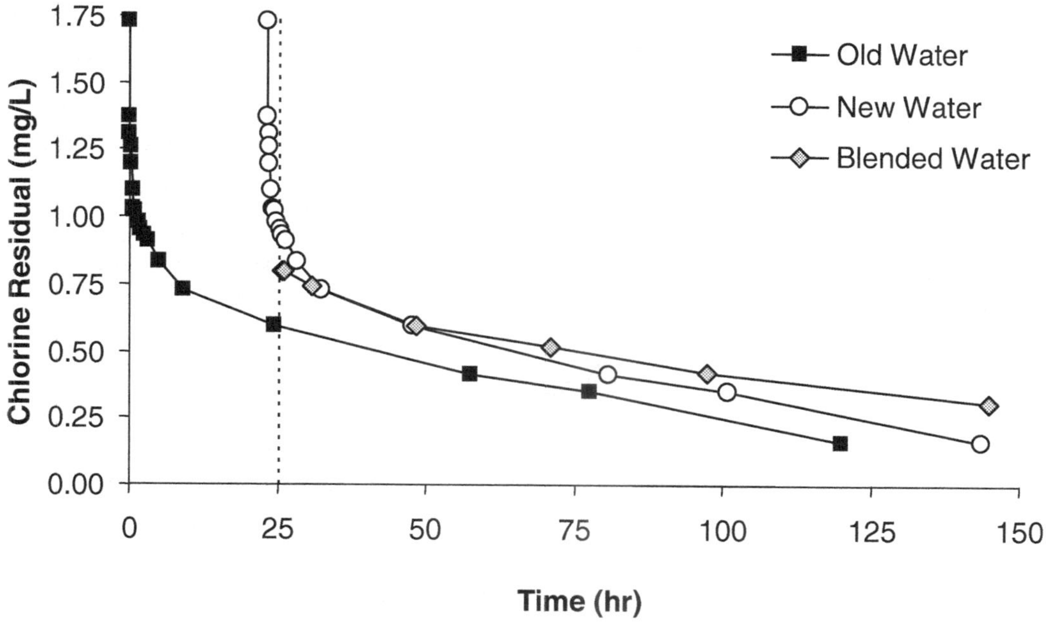


**Figure 3.13 Effect of blending 24-hour chlorinated "old water" with 2-hour chlorinated "new water" at a 1:1 volume ratio. Betasso, GAC, TOC = 1.15 mg/L, 1.5:1 $Cl_2$:TOC dose (mg/mg), pH 8.0, 20°C.**

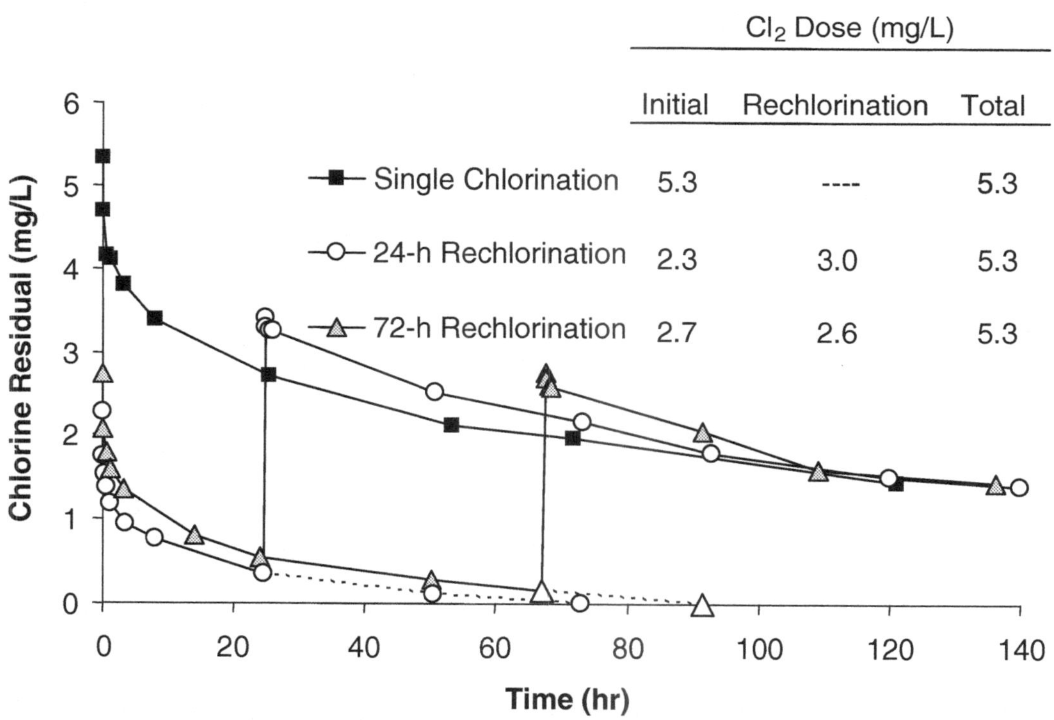


**Figure 3.14 Effect of rechlorination on chlorine decay. NPNV, EC, TOC = 3.56 mg/L, 1.5:1 $Cl_2$:TOC dose (mg/mg), pH 8.0, 20°C.**

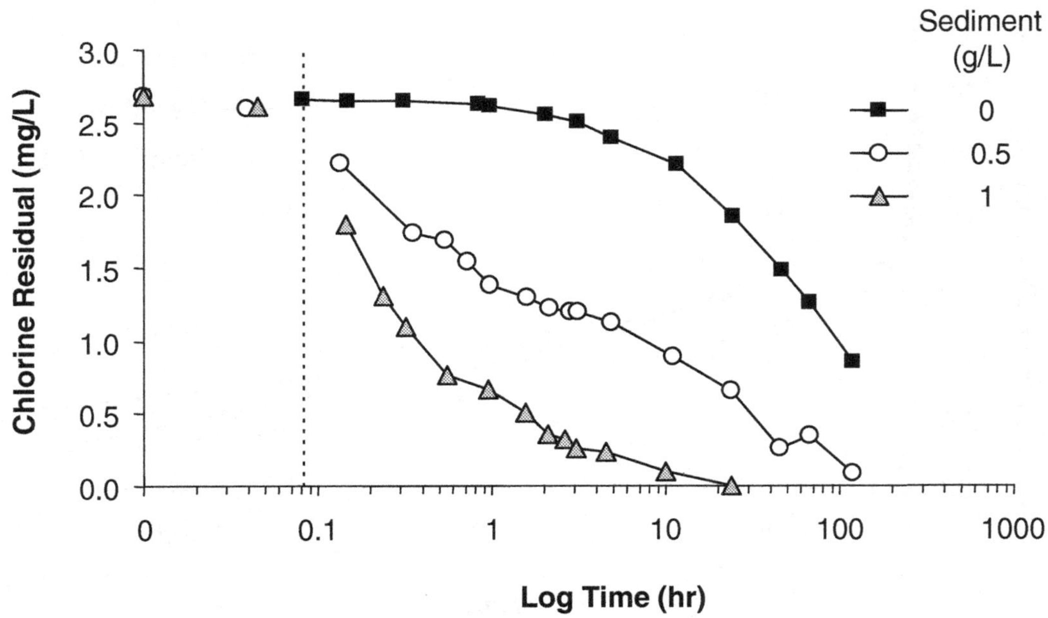


**Figure 3.15 Effect of ductile iron pipe sediment on chlorine decay. Boulder tap water, TOC = 0.99 mg/L, $Cl_2$ residual = 0.67 mg/L, rechlorinated with an additional 2.0 mg/L. Sediment added five minutes after rechlorination.**

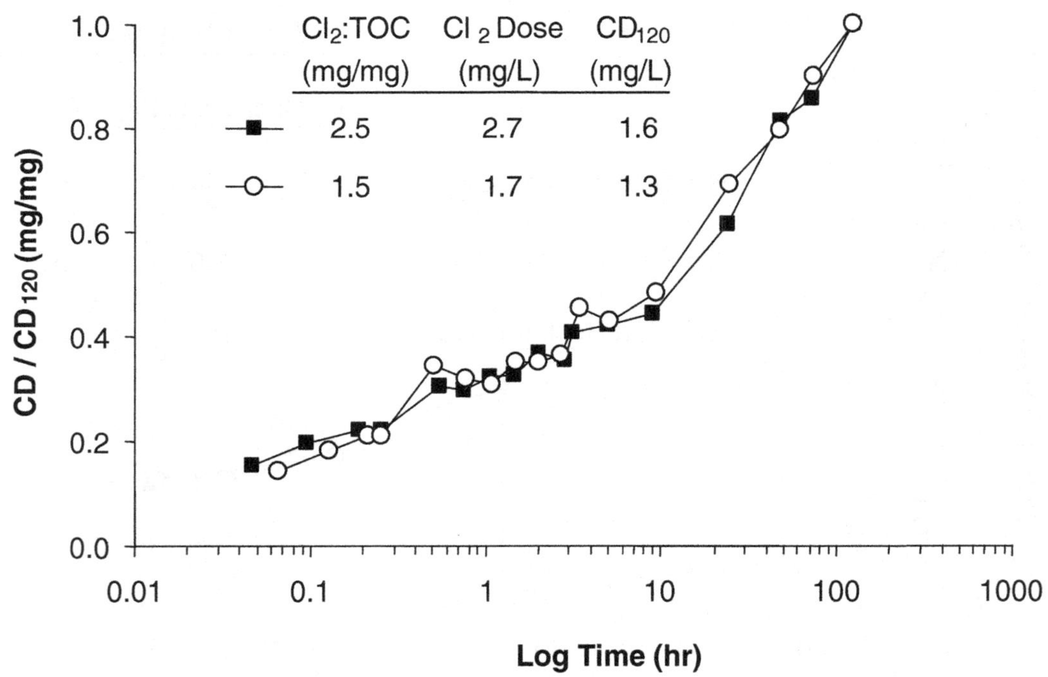


**Figure 3.16 Effect of chlorine dose on chlorine consumption kinetics normalized by 120-hour chlorine demand. Betasso, $O_3$/BF, TOC = 1.10 mg/L, pH 8.0, 20°C.**

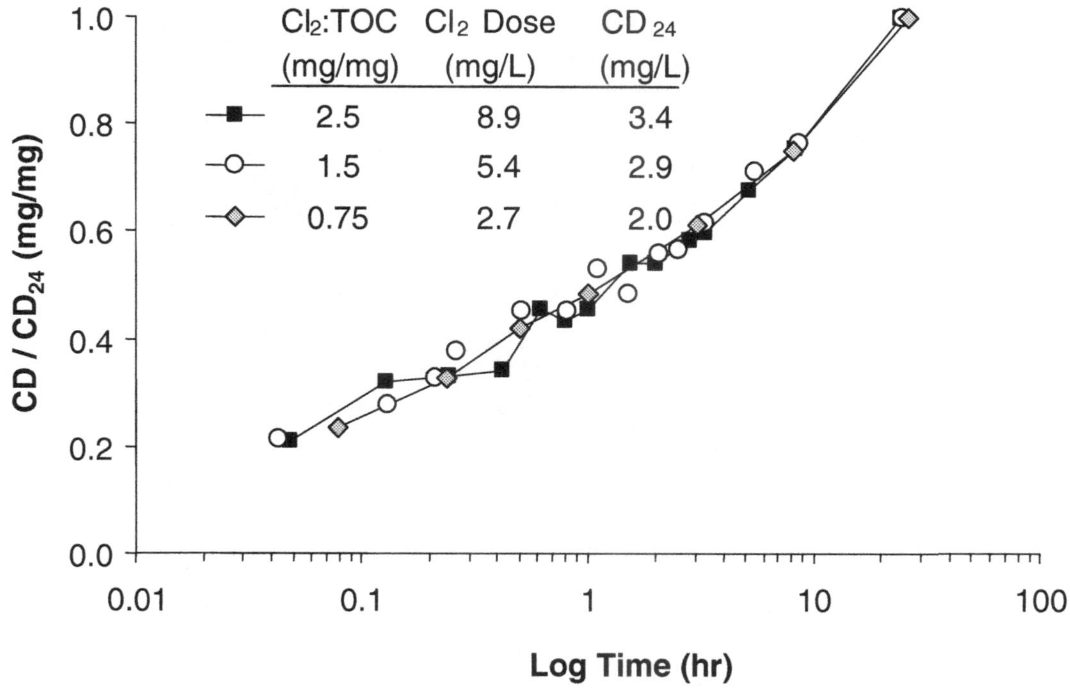


**Figure 3.17 Effect of chlorine dose on chlorine consumption kinetics normalized by 24-hour chlorine demand. NPNV, EC, TOC = 3.56 mg/L, pH 8.0, 20°C.**

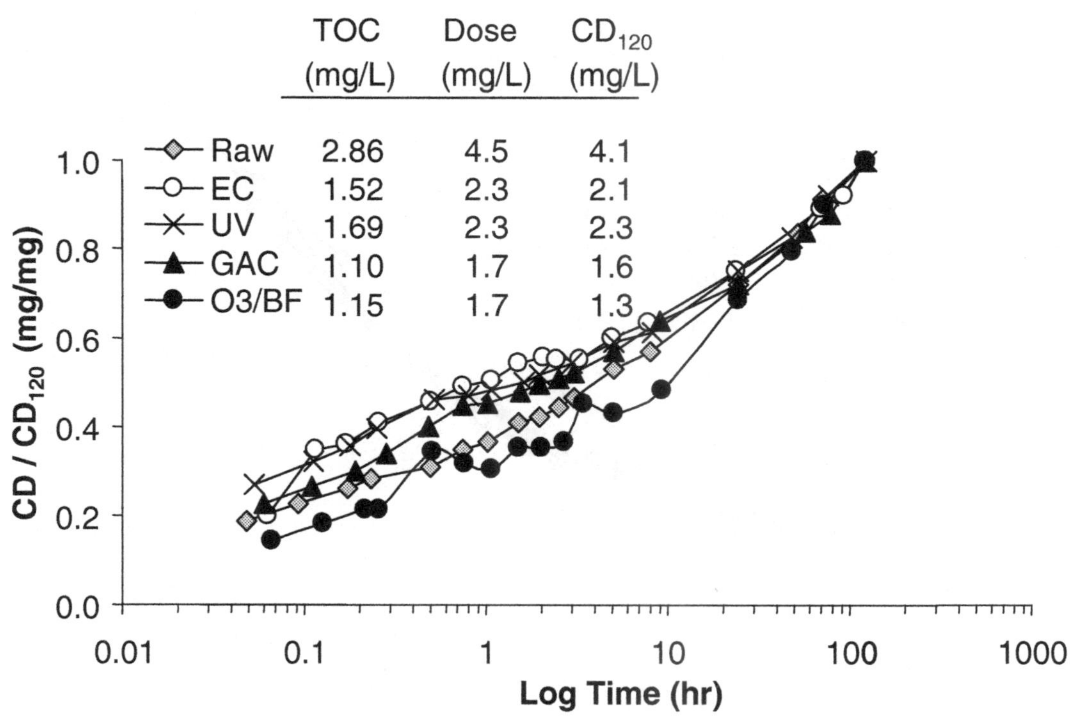


**Figure 3.18 Effect of treatment on normalized chlorine demand kinetics. Betasso, 1.5:1 $Cl_2$:TOC dose (mg/mg), pH 8.0, 20°C.**

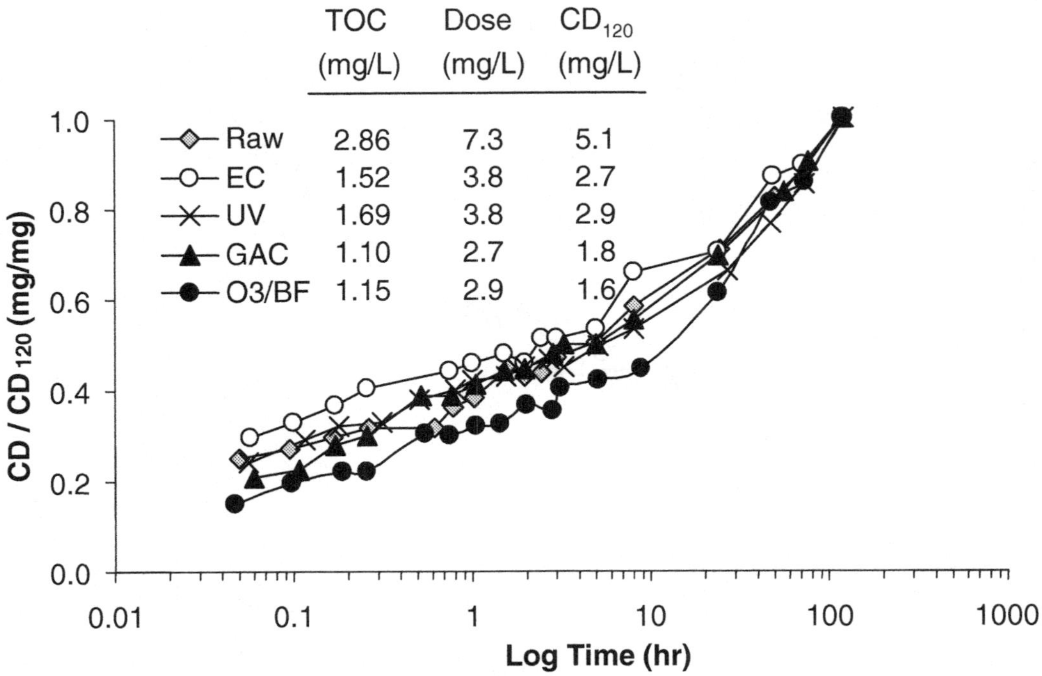


**Figure 3.19 Effect of treatment on normalized chlorine demand kinetics. Betasso, 2.5:1 $Cl_2$:TOC dose (mg/mg), pH 8.0, 20°C.**

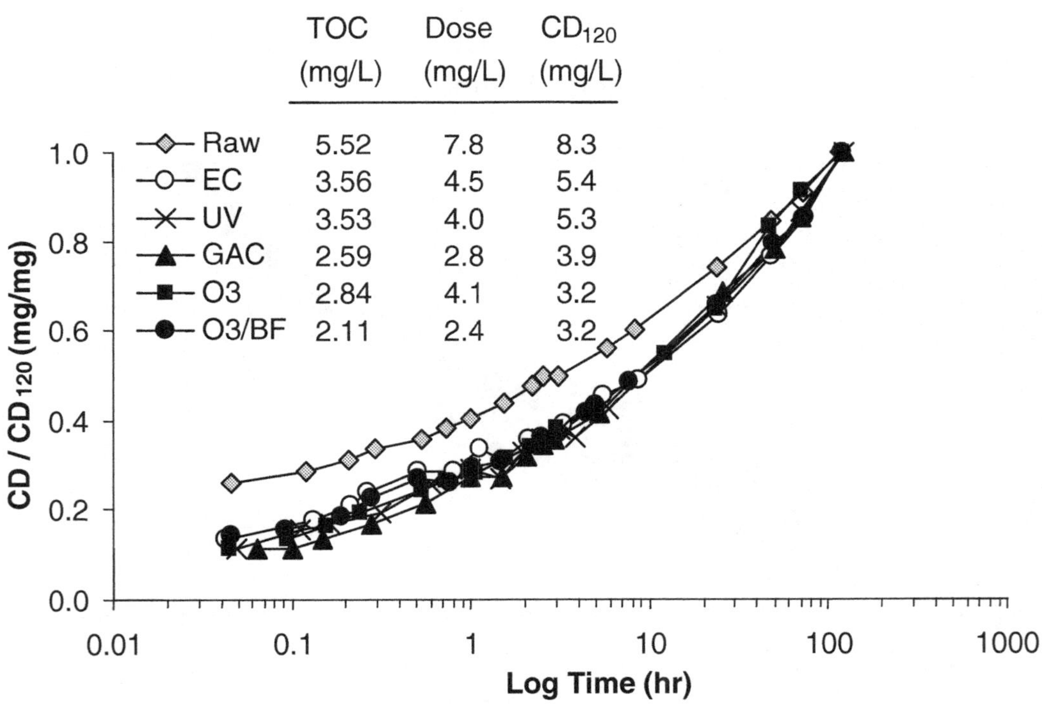


**Figure 3.20 Effect of treatment on normalized chlorine demand kinetics. NPNV, 1.5:1 $Cl_2$:TOC dose (mg/mg), pH 8.0, 20°C.**

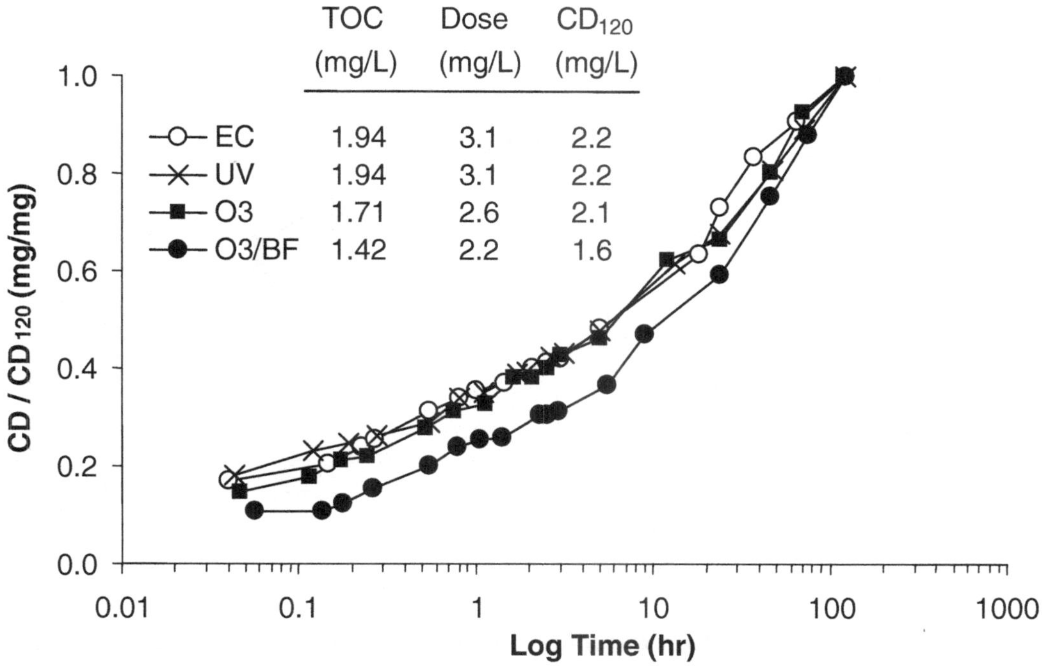


**Figure 3.21 Effect of treatment on normalized chlorine demand kinetics. Winchester, 1.5:1 $Cl_2$:TOC dose (mg/mg), pH 8.0, 20°C.**

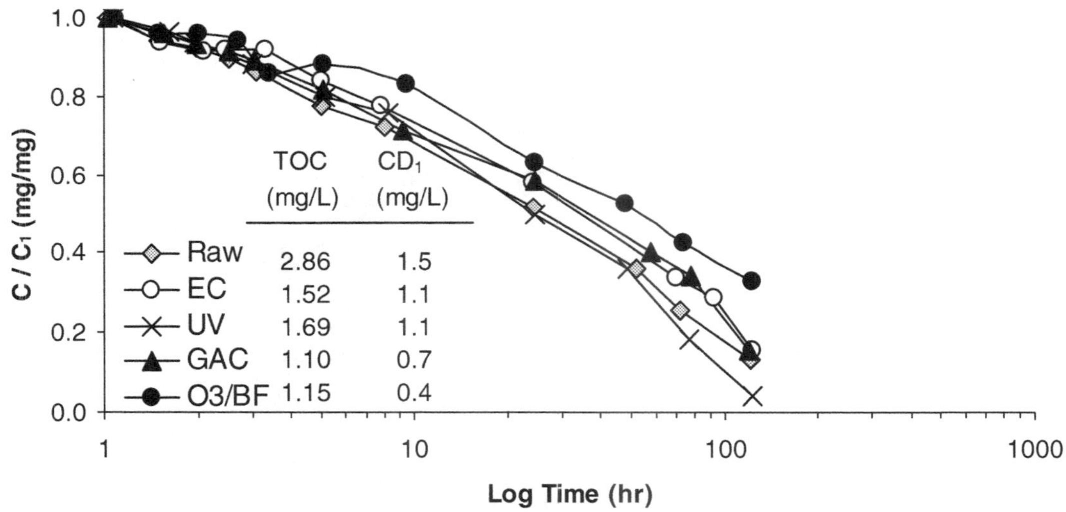


**Figure 3.22 Effect of treatment on chlorine decay normalized by 1-hour residual. Betasso, 1.5:1 $Cl_2$:TOC dose (mg/mg), pH 8.0, 20°C.**

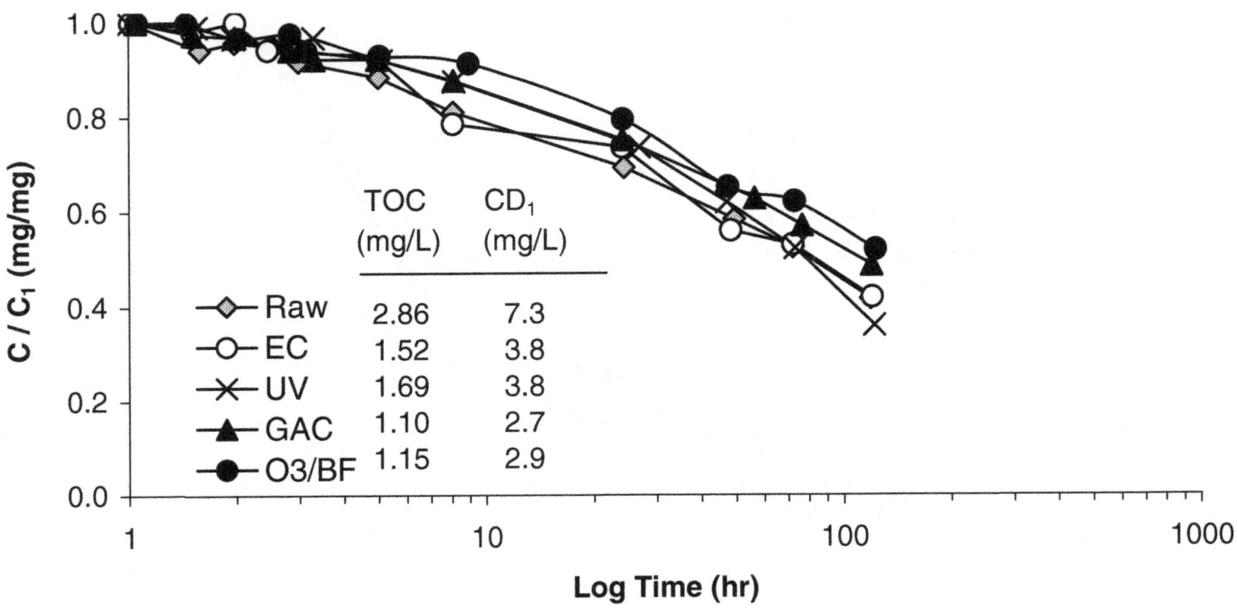


**Figure 3.23 Effect of treatment on chlorine decay normalized by 1-hour residual. Betasso, 2.5:1 $Cl_2$:TOC dose (mg/mg), pH 8.0, 20°C.**

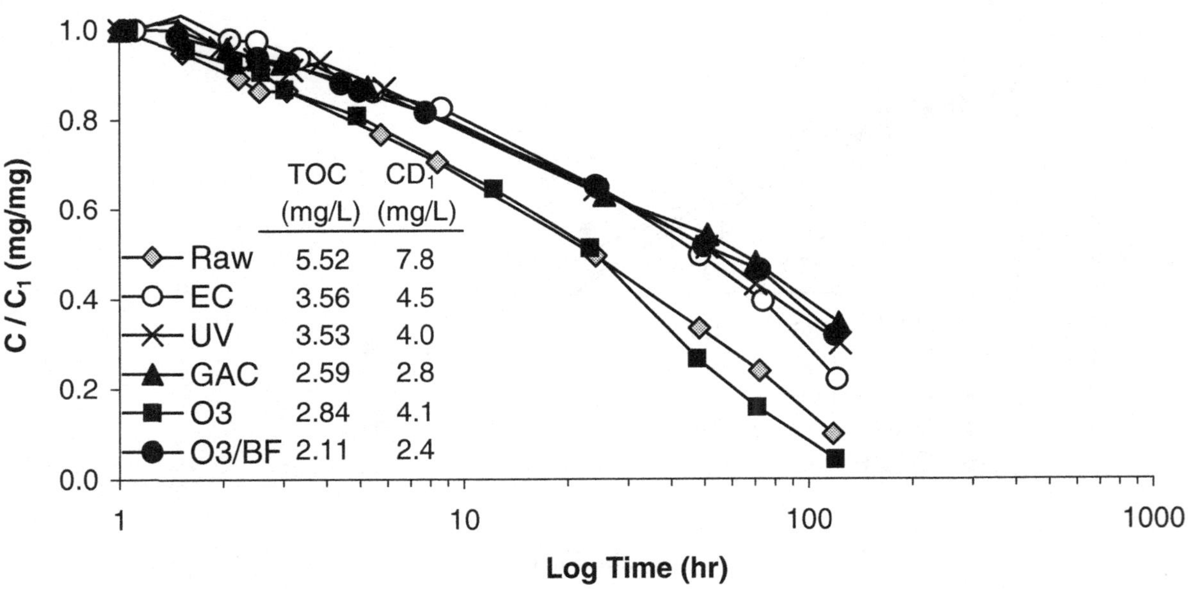


**Figure 3.24 Effect of treatment on chlorine decay normalized by 1-hour residual. NPNV, 1.5:1 $Cl_2$:TOC dose (mg/mg), pH 8.0, 20°C.**

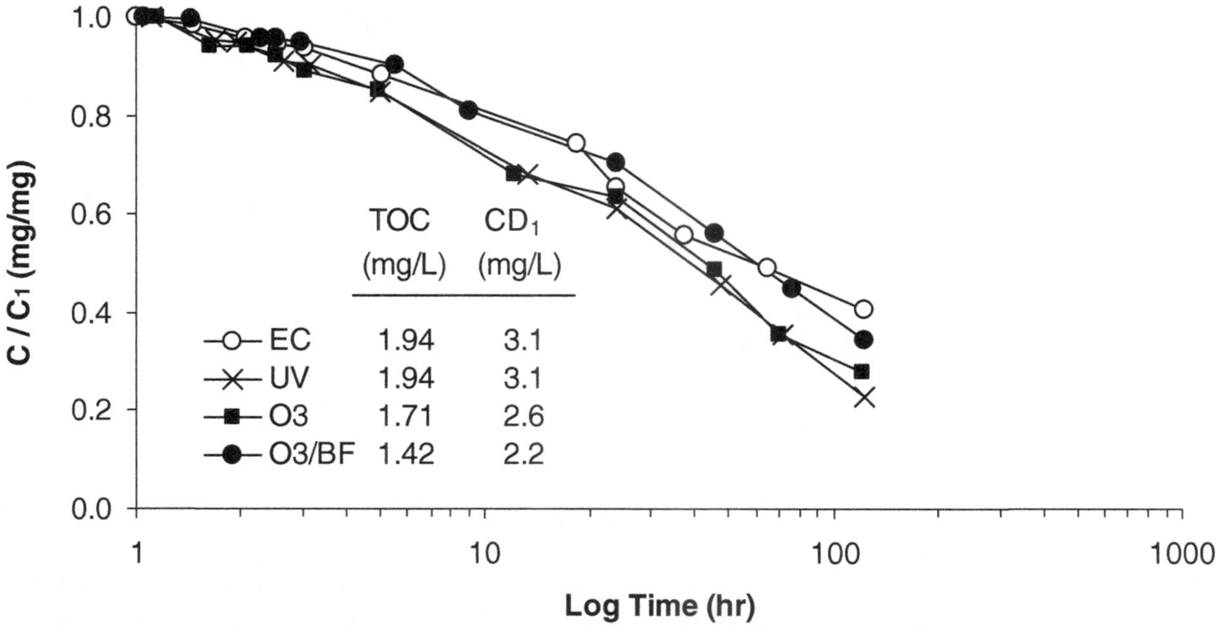


**Figure 3.25 Effect of treatment on chlorine decay normalized by 1-hour residual. Winchester, 1.5:1 $Cl_2$:TOC dose (mg/mg), pH 8.0, 20°C.**

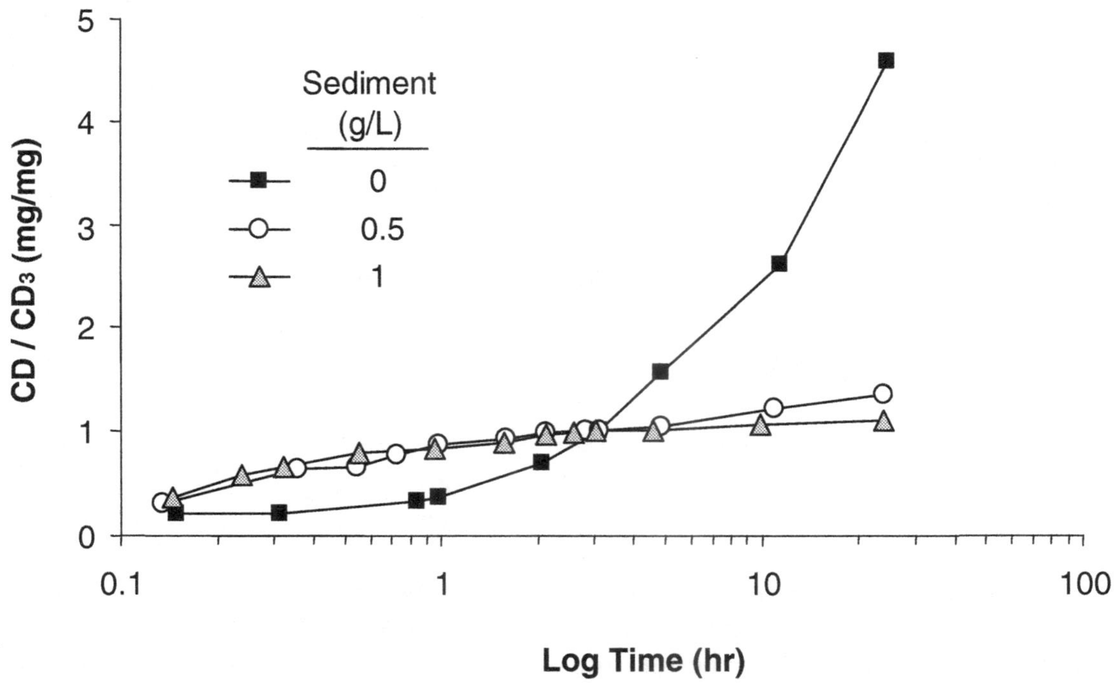


**Figure 3.26 Effect of ductile iron pipe sediment on chlorine consumption normalized by the 3-hour chlorine demand. Boulder tap water, TOC = 0.99 mg/L, $Cl_2$ residual = 0.67 mg/L, rechlorinated with an additional 2.0 mg/L. Sediment added five minutes after rechlorination. pH 7.5 – 8.0, 20°C.**

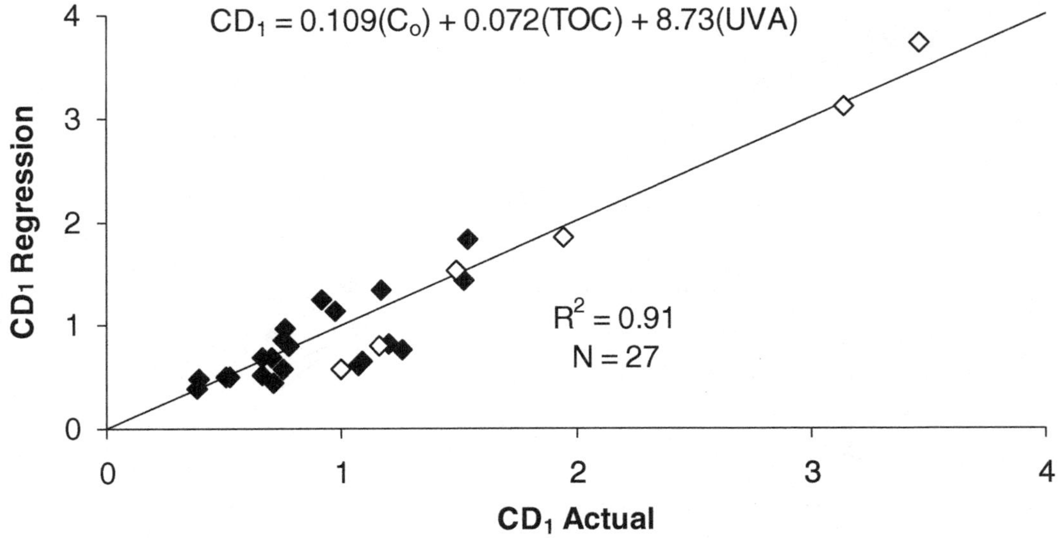


**Figure 3.27 Regression vs. actual 1-hour chlorine demands. Solid symbols indicate values used for regression. Hollow symbols (2 Betasso Raw, 2 NPNV Raw, 2 NPNV $O_3$) not used for regression. Betasso, NPNV, Winchester source waters; EC, UV, GAC, $O_3$, $O_3$/BF; 0.75:1, 1.5:1, 2.5:1 $Cl_2$:TOC dose (mg/mg), pH 8.0, 20°C.**

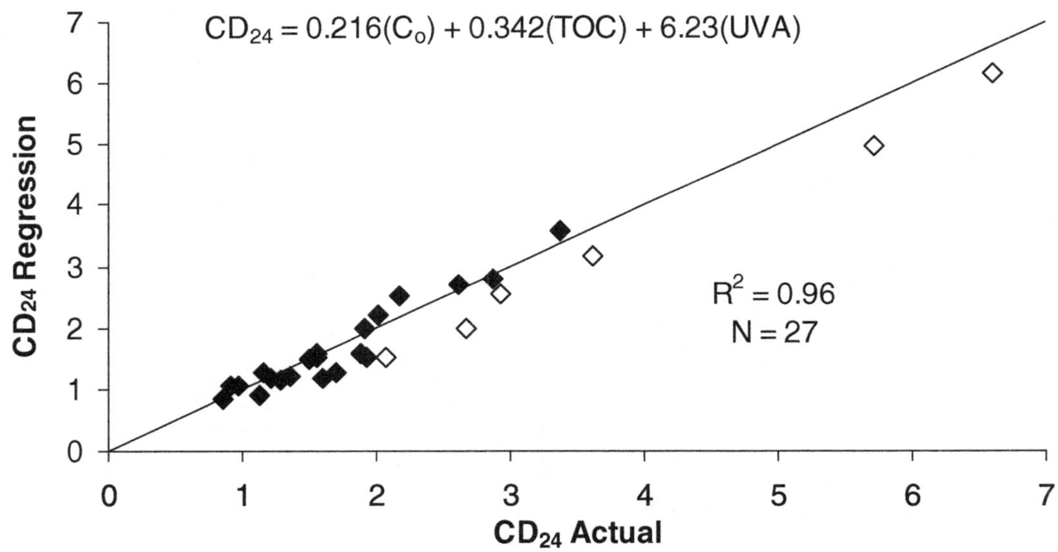


**Figure 3.28 Regression vs. actual 24-hour chlorine demands. Solid symbols indicate values used for regression. Hollow symbols (2 Betasso Raw, 2 NPNV Raw, 2 NPNV $O_3$) not used for regression. Betasso, NPNV, Winchester source waters; EC, UV, GAC, $O_3$, $O_3$/BF; 0.75:1, 1.5:1, 2.5:1 $Cl_2$:TOC dose (mg/mg), pH 8.0, 20°C.**

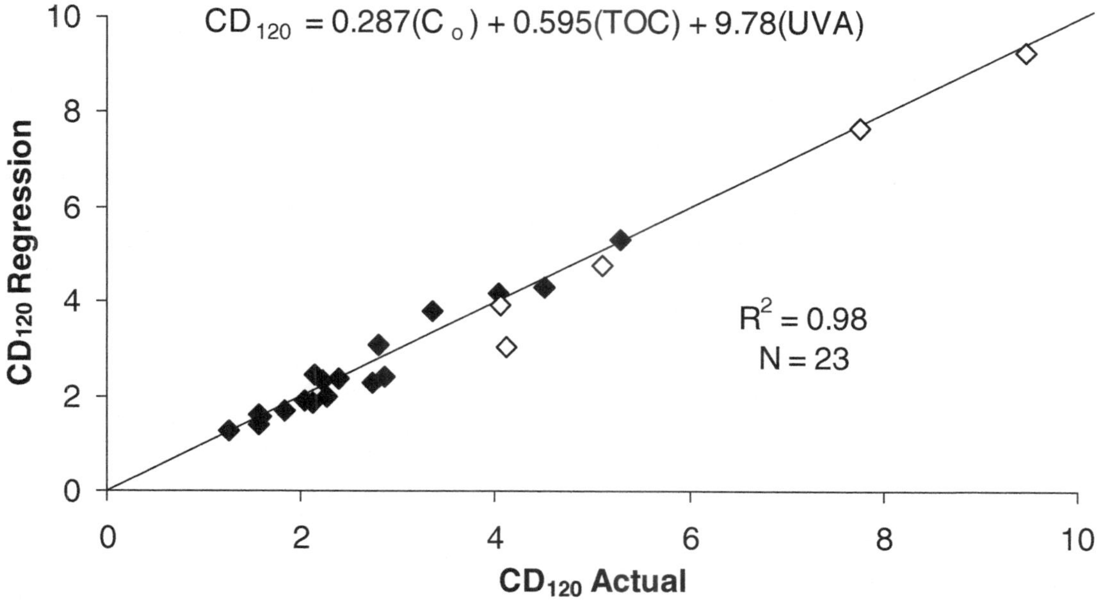


**Figure 3.29 Regression vs. actual 120-hour chlorine demands. Solid symbols: values used for regression. Hollow symbols (2 Betasso Raw, 2 NPNV Raw, 2 NPNV $O_3$): not used for regression. Betasso, NPNV, Winchester source waters; EC, UV, GAC, $O_3$, $O_3$/BF; 1.5:1, 2.5:1 $Cl_2$:TOC dose (mg/mg), pH 8.0, 20°C.**

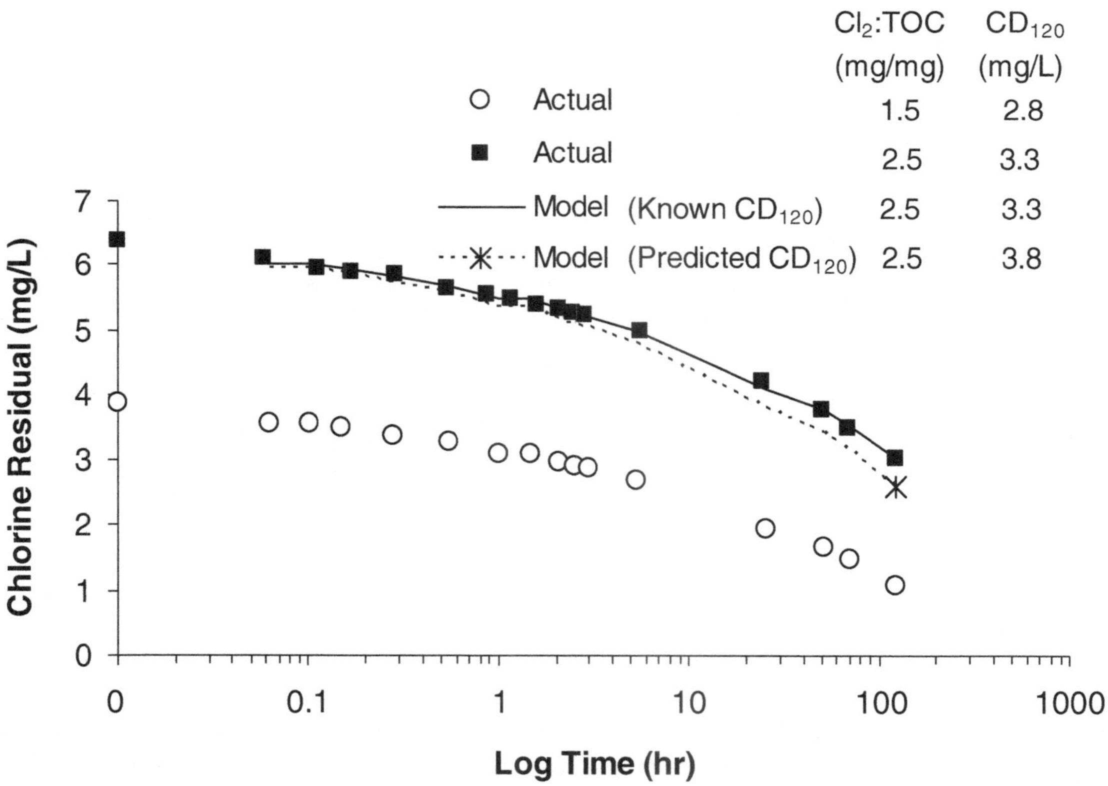


**Figure 3.30 Predicting the chlorine residual curve at 2.5:1 $Cl_2$:TOC dose from the known chlorine decay curve at 1.5:1 $Cl_2$:TOC dose, from both known $CD_{120}$ and predicted $CD_{120}$ (Linear Regression Model). Betasso, GAC, pH 8.0, 20°C.**

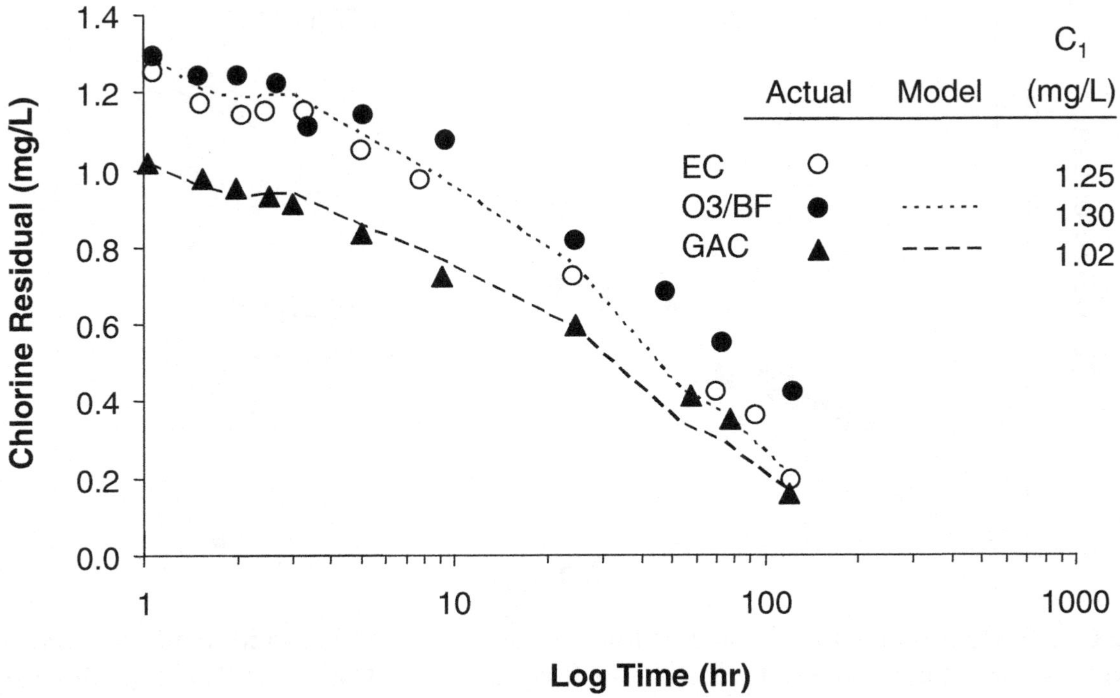


**Figure 3.31 Predicting the chlorine residual curve for GAC and $O_3$/BF treated waters from the known EC water chlorine decay curve and the known $C_1$ for each water. Betasso, 1.5:1 $Cl_2$:TOC dose, pH 8.0, 20°C.**

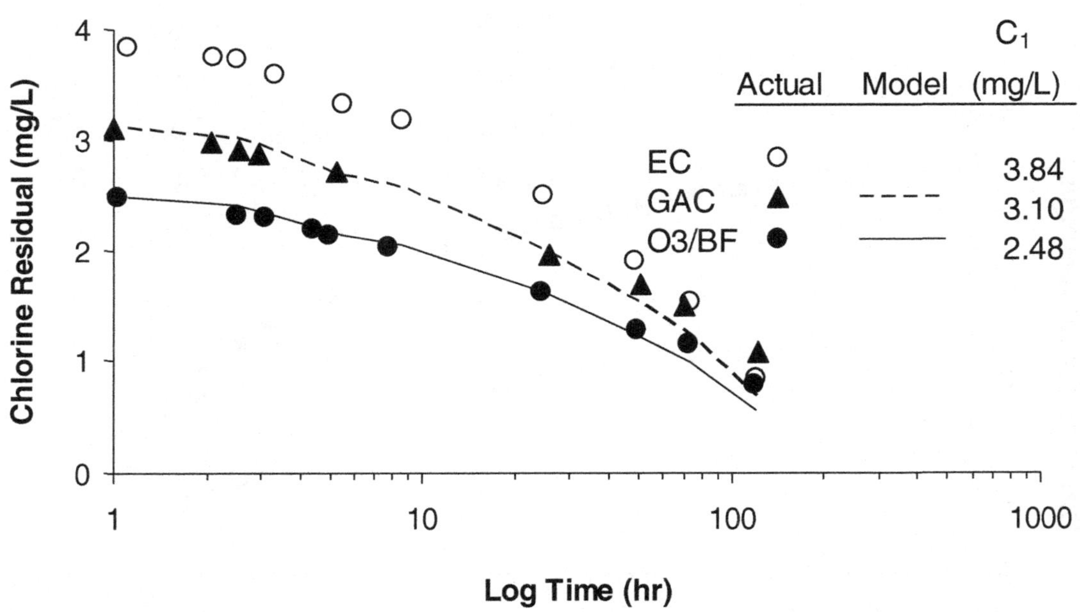


**Figure 3.32 Predicting the chlorine residual curve for GAC and $O_3$/BF treated waters from the known EC water chlorine decay curve and the known $C_1$ for each water. NPNV, 1.5:1 $Cl_2$:TOC dose, pH 8.0, 20°C.**

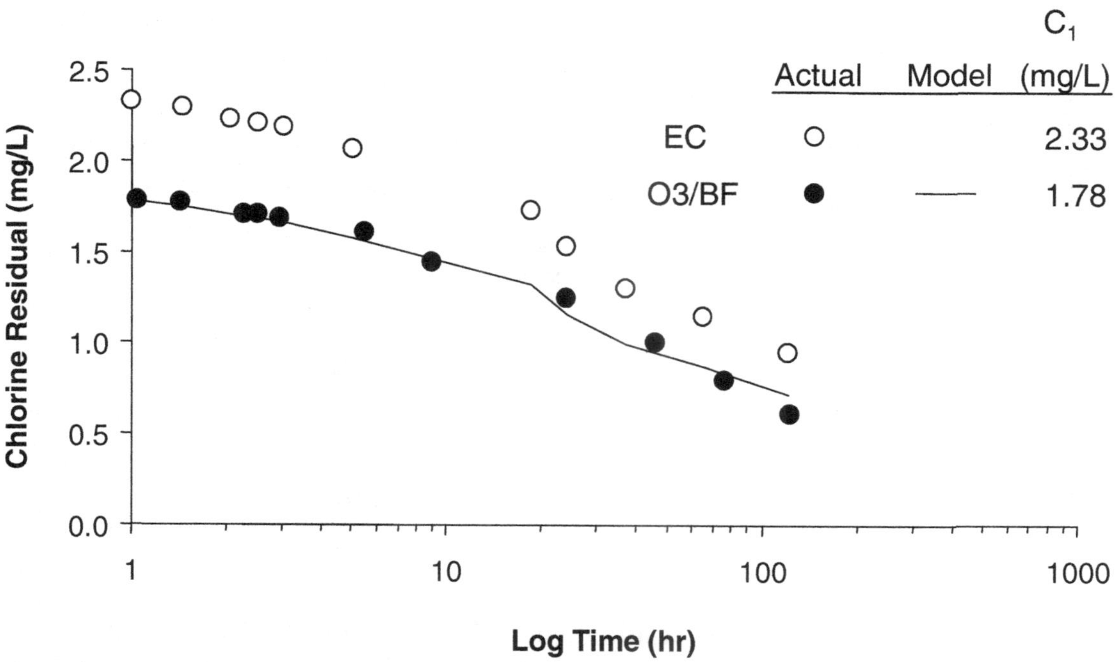


**Figure 3.33 Predicting the chlorine residual curve for $O_3$/BF treated water from the known EC water chlorine decay curve and the known $C_1$ for $O_3$/BF water. Winchester, 1.5:1 $Cl_2$:TOC dose, pH 8.0, 20°C.**

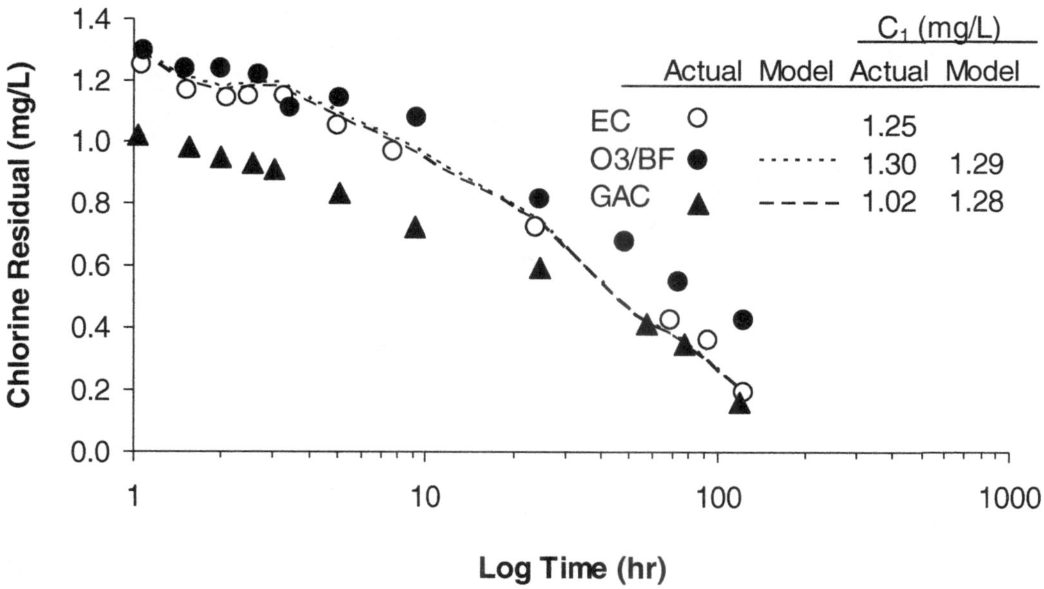


**Figure 3.34 Predicting the chlorine residual curve for GAC and $O_3$/BF treated waters from the known EC water chlorine decay curve and the predicted $C_1$ for each water ($CD_1$ Linear Regression Model). Betasso, 1.5:1 $Cl_2$:TOC dose, pH 8.0, 20°C.**

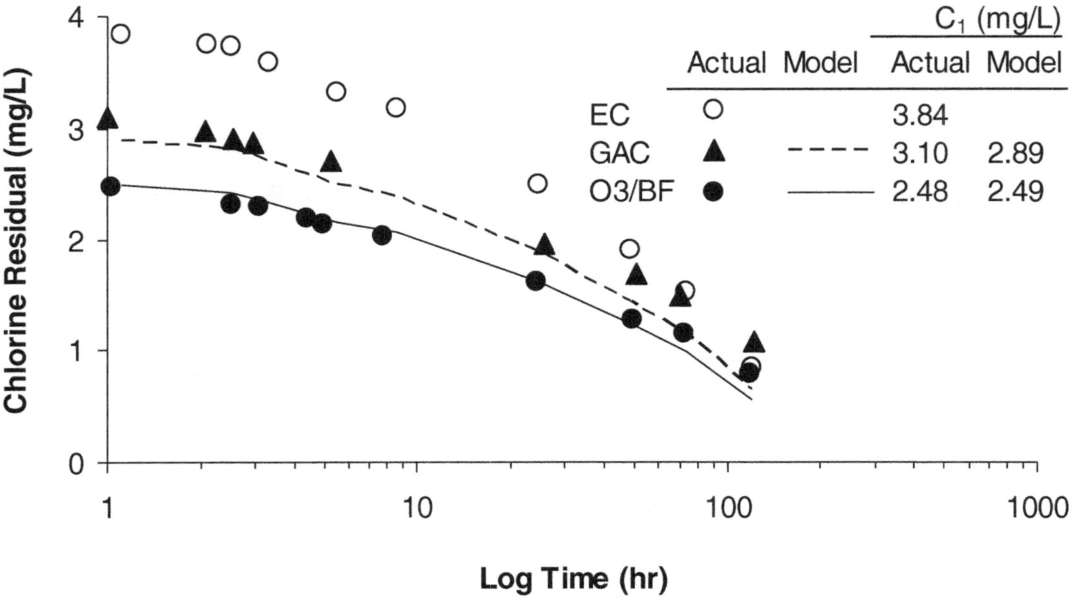


**Figure 3.35 Predicting the chlorine residual curve for GAC and $O_3$/BF treated waters from the known EC water chlorine decay curve and the predicted $C_1$ for each water ($CD_1$ Linear Regression Model). NPNV, 1.5:1 $Cl_2$:TOC dose, pH 8.0, 20°C.**

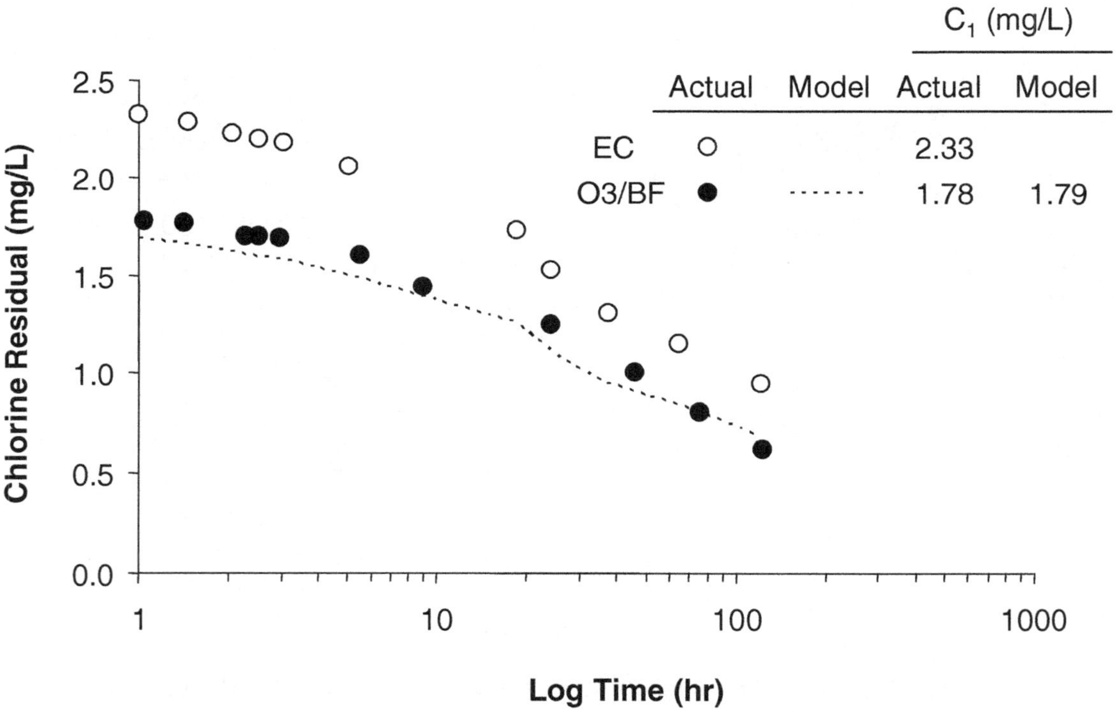


**Figure 3.36 Predicting the chlorine residual curve for $O_3$/BF treated water from the known EC water chlorine decay curve and the predicted $C_1$ for $O_3$/BF Water ($CD_1$ Linear Regression Model). Winchester, 1.5:1 $Cl_2$:TOC Dose, pH 8.0, 20°C.**

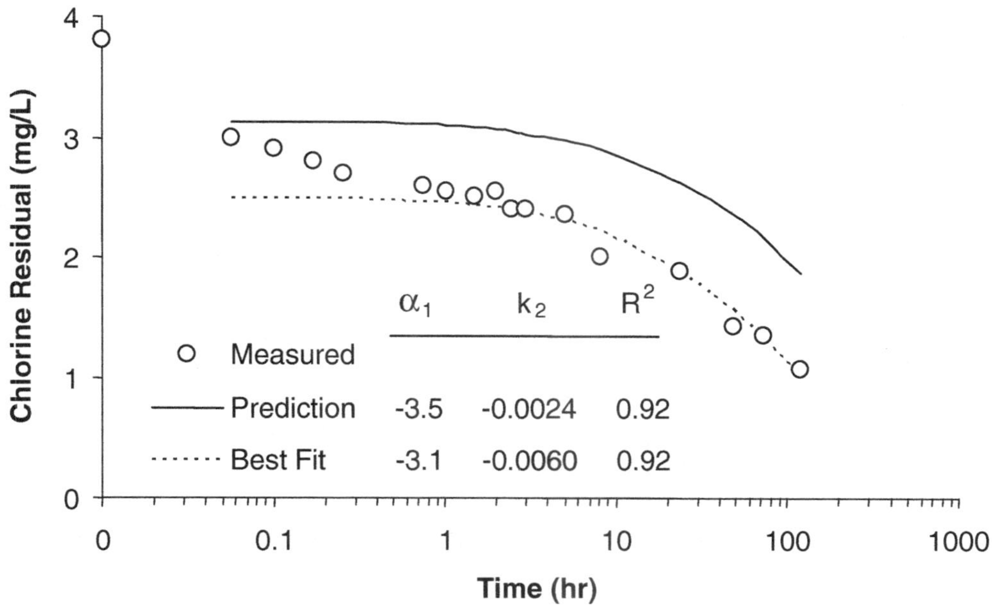


**Figure 3.37 Koechling (1998)/Isabel (2000) model of Betasso, EC. TOC = 1.52 mg/L, 2.5:1 $Cl_2$:TOC dose, pH 8.0, 20°C.**

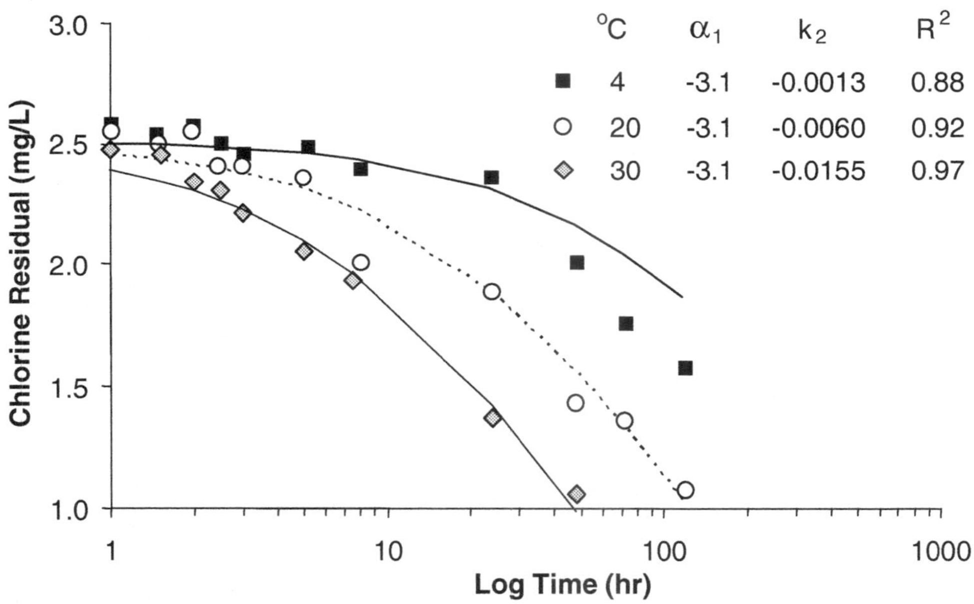


**Figure 3.38 Predicting the change in rate constant, $k_2$, for Koechling (1998) model with change in temperature. Betasso, EC, TOC = 1.52 mg/L, 2.5:1 $Cl_2$:TOC dose, pH 8.0.**

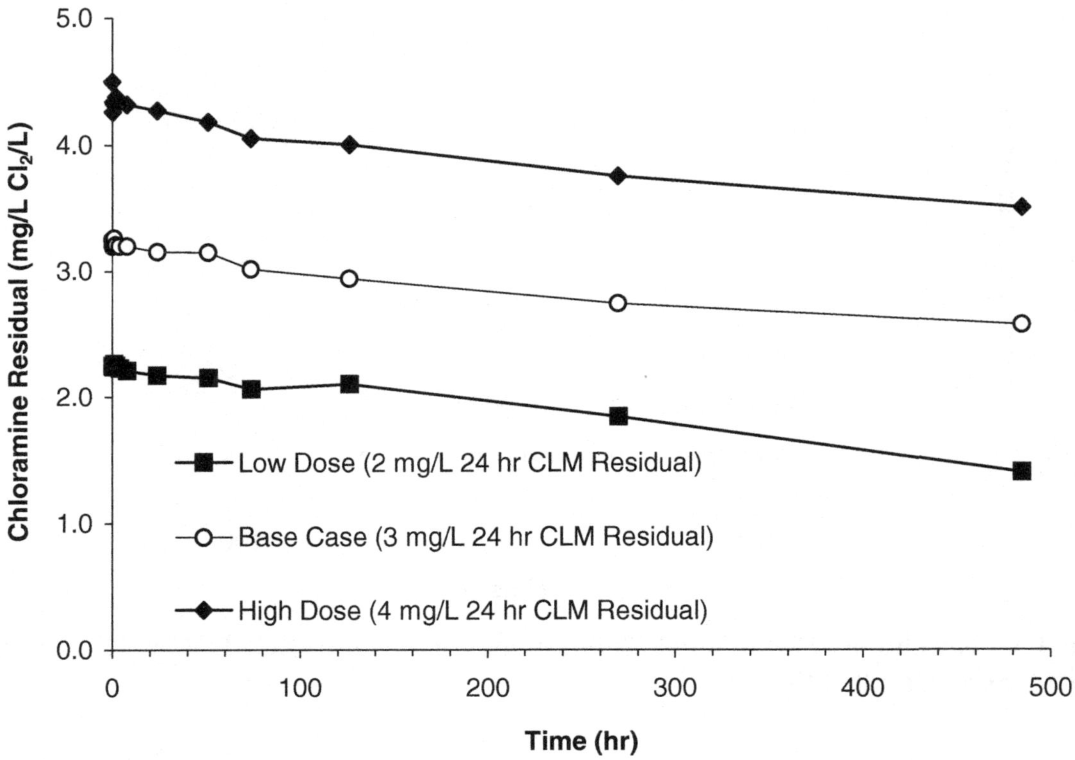


**Figure 3.39 Effect of dose on chloramine residual. $O_3$/BF Betasso, pH 8.0, 20°C, TOC 1.1 mg/L.**

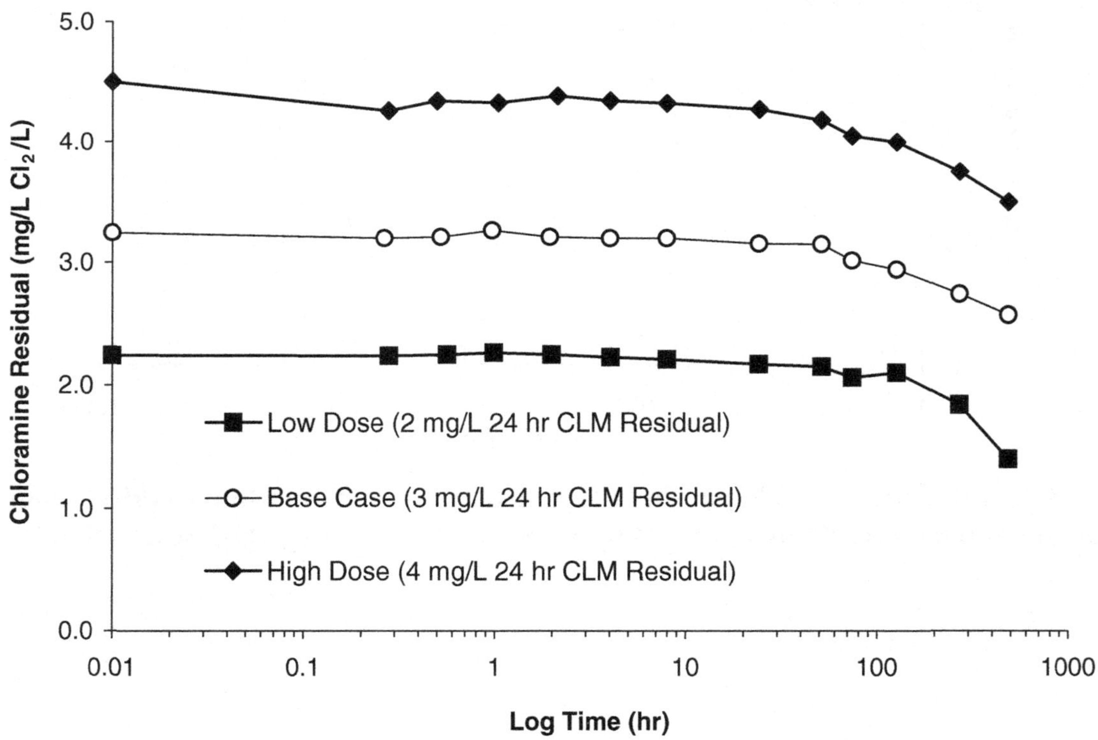


**Figure 3.40 Effect of dose on chloramine decay. $O_3$/BF Betasso, pH 8.0, 20°C.**

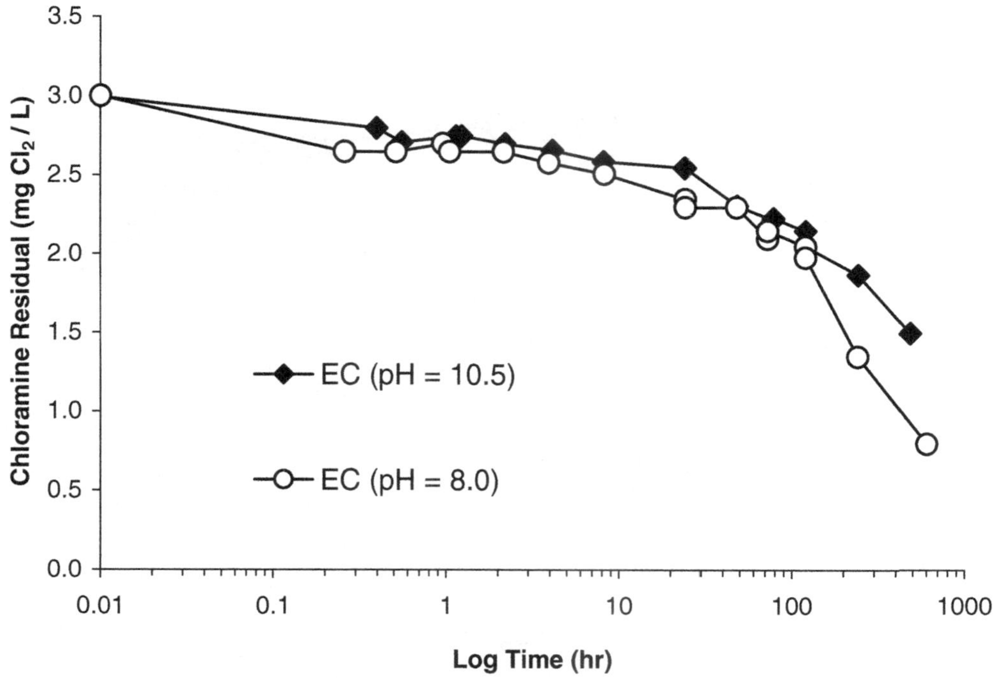


**Figure 3.41 Effect of pH on chloramine decay. EC NPNV, pH 8.0, 20°C, 2 mg/L chloramine residual at 24 hrs.**

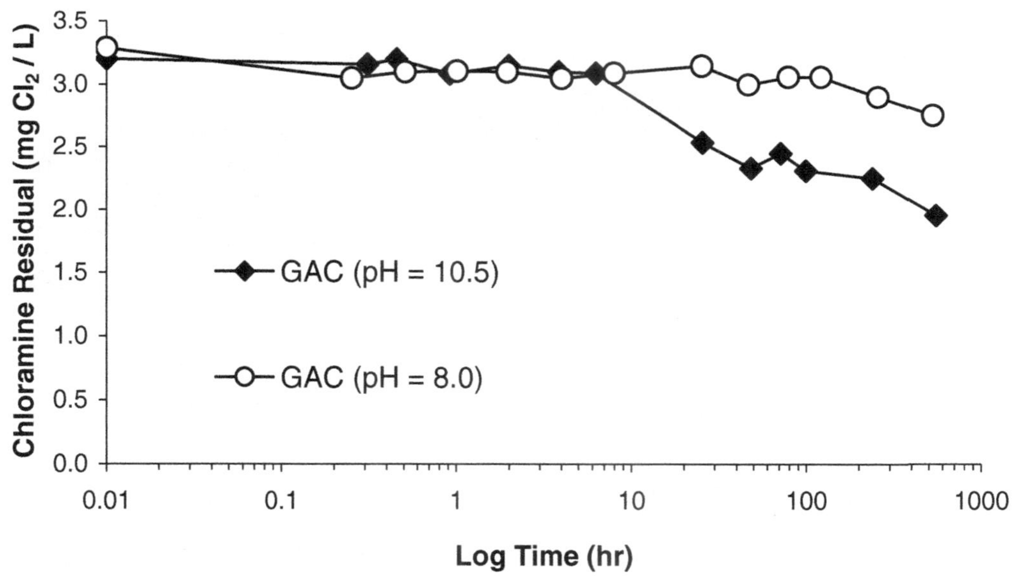


**Figure 3.42 Effect of pH on chloramine decay. GAC Betasso, pH 8.0, 20°C, 2 mg/L chloramine residual at 24 hrs.**

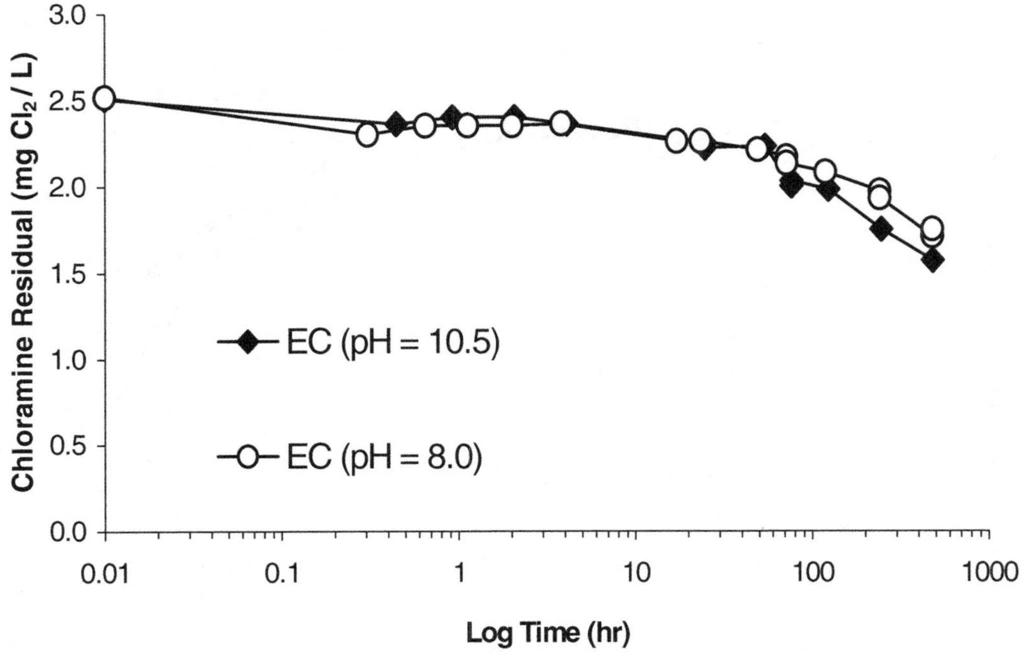


**Figure 3.43 Effect of pH on chloramine decay. EC Winchester, pH 8.0, 20°C, 2 mg/L chloramine residual at 24 hrs.**

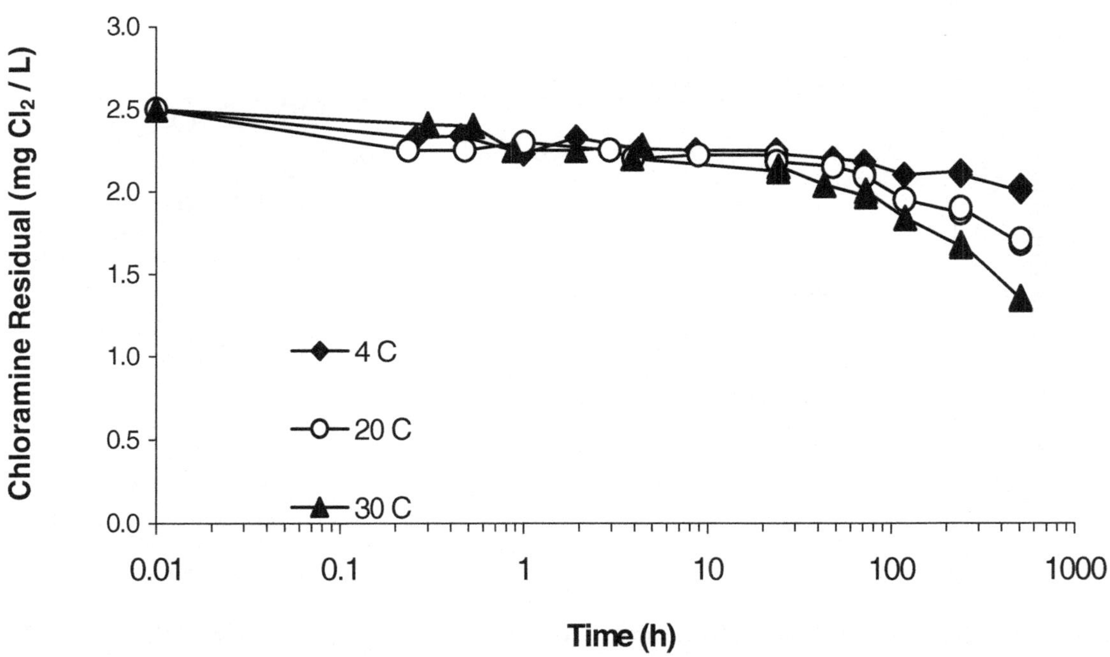


**Figure 3.44 Effect of temperature on chloramine decay. $O_3$/BF Winchester, pH 8.0, 2 mg/L chloramine residual at 24 hrs.**

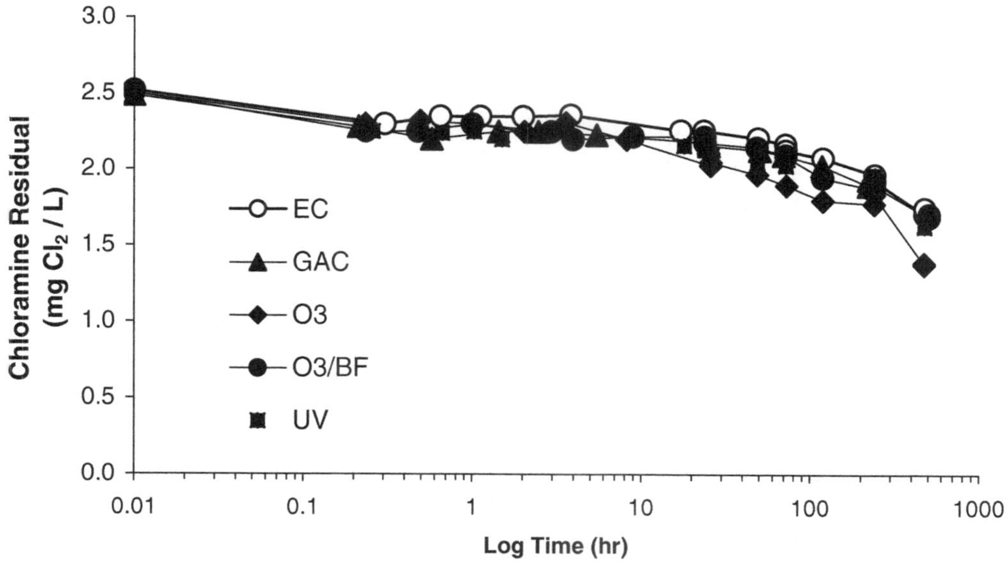


**Figure 3.45 Effect of treatment on chloramine decay. Winchester, pH 8.0, 20°C, 2 mg/L chloramine residual at 24 hrs.**

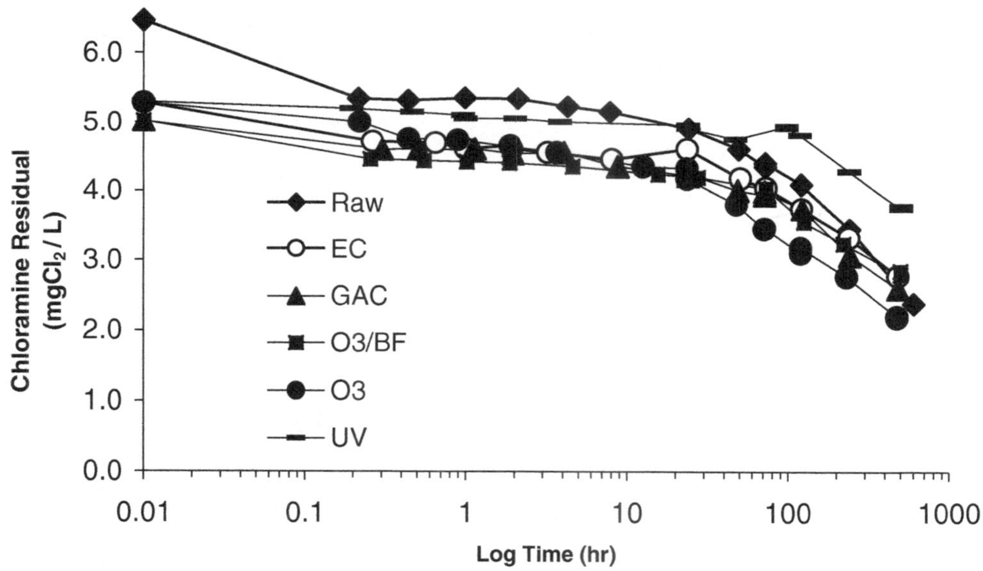


**Figure 3.46 Effect of treatment on chloramine decay. NPNV, pH 8.0, 20°C. 4 mg/L chloramine residual at 24 hrs.**

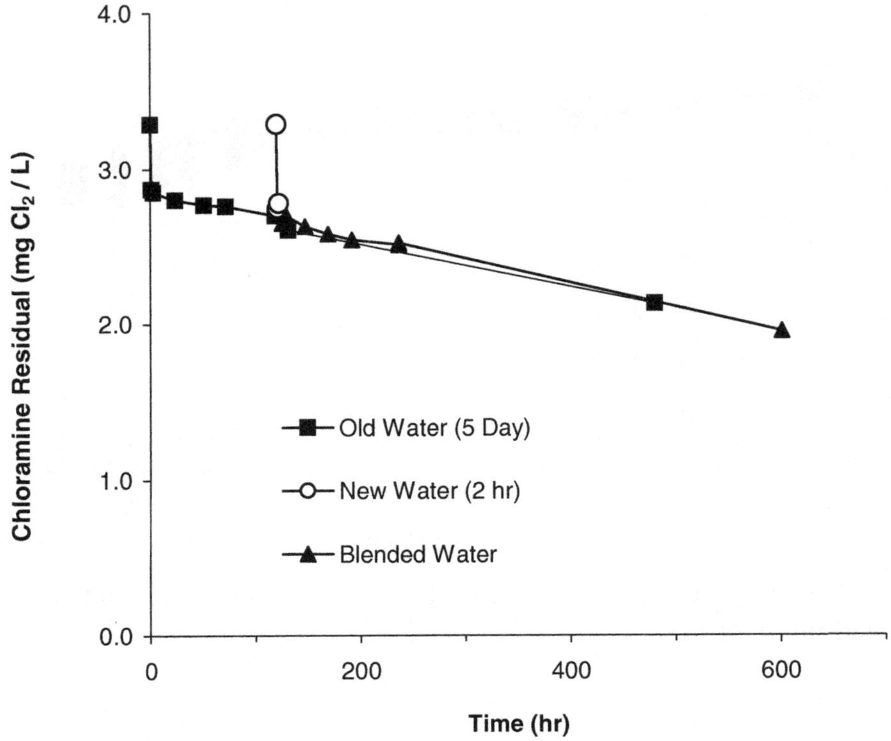


**Figure 3.47 Effect of water age on blended chloramine decay. Waters reacted for 5 days and 2 hours. GAC Betasso, pH 8.0, 20°C.**

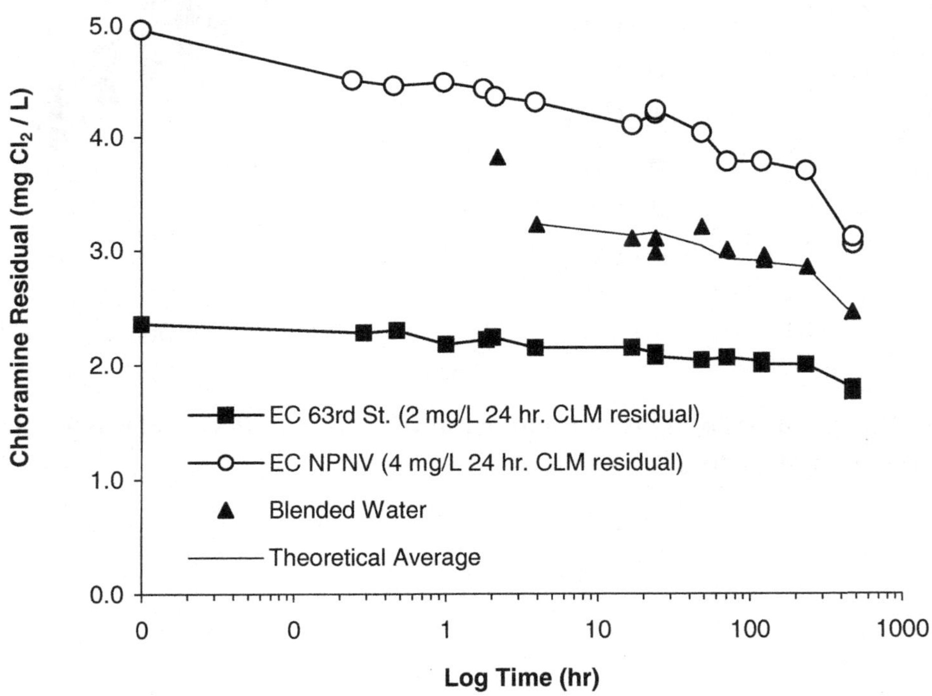


**Figure 3.48 Effect of residual on blended chloramine decay. NPNV, pH 8.0, 20°C.**

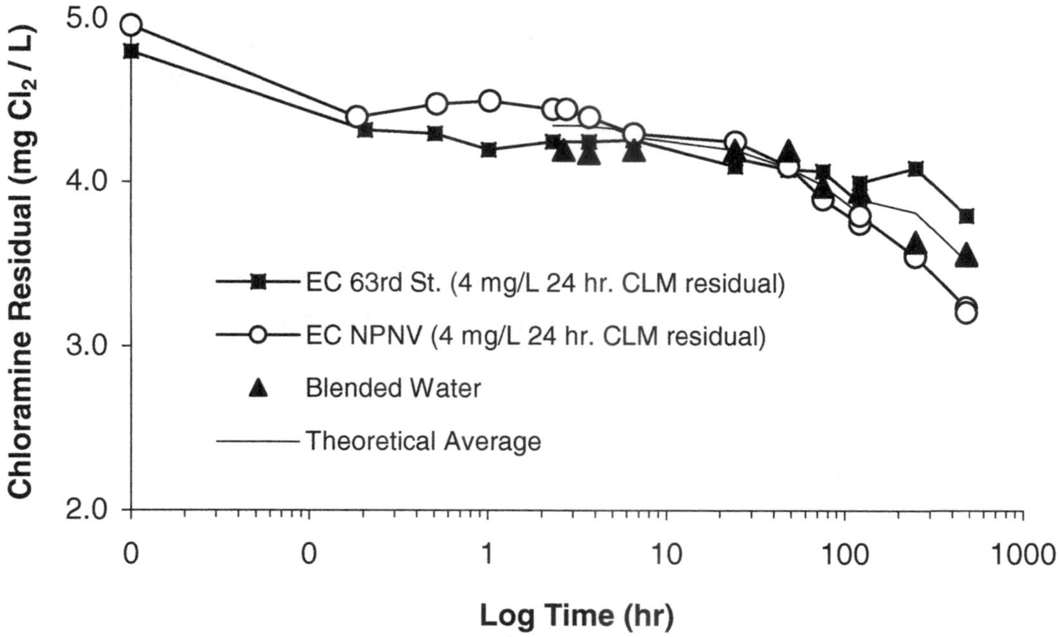


**Figure 3.49 Effect of demand on blended chloramine decay. pH 8.0, 20°C, EC NPNV TOC 3.6 mg/L, EC 63rd Street TOC 1.9 mg/L.**

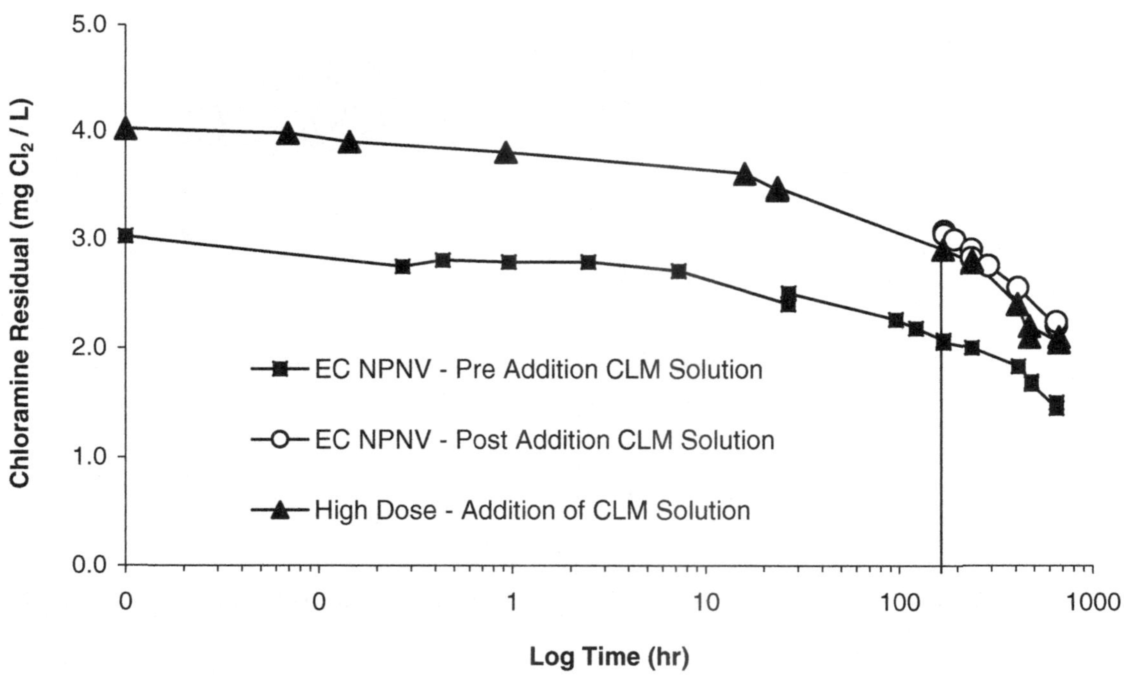


**Figure 3.50 Effect of rechloramination on chloramine decay. EC Winchester, pH 8.0, 20°C.**

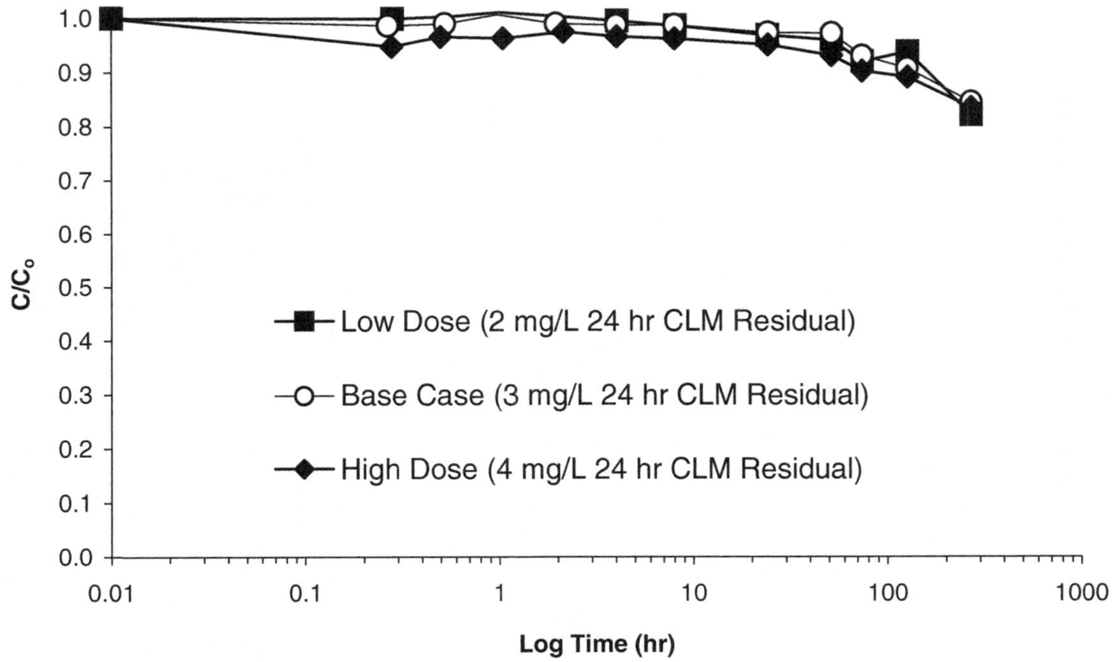


**Figure 3.51 Effect of dose on normalized chloramine residual. $O_3$/BF Betasso, 20°C, pH 8.0, TOC 1.1 mg/L.**

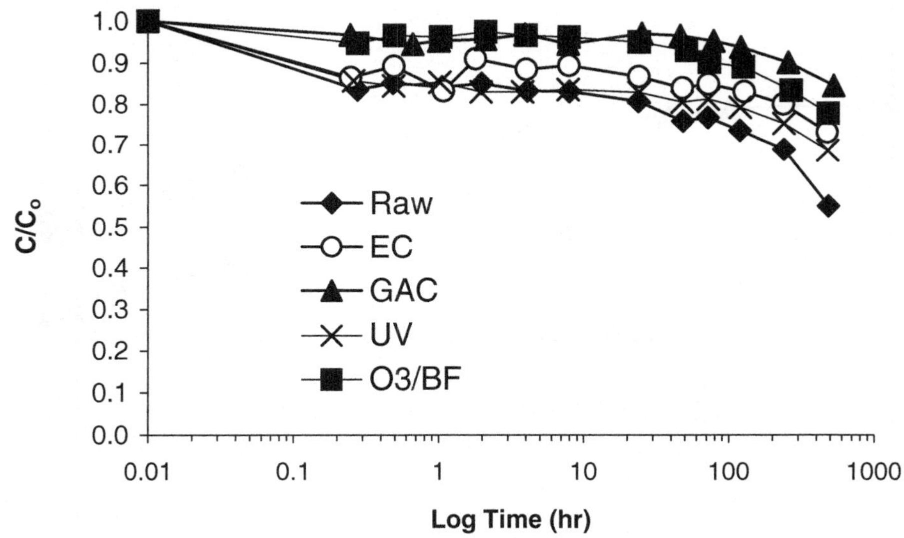


**Figure 3.52 Effect of treatment on normalized chloramine residual. Betasso, pH 8.0, 20°C, 4 mg/L chloramine residual at 24 hrs.**

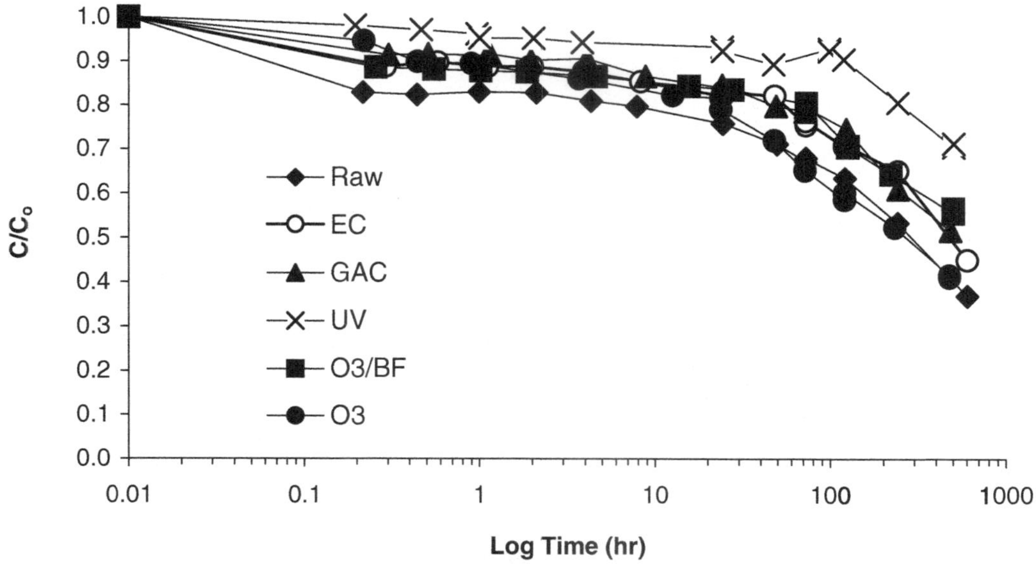


**Figure 3.53 Effect of treatment on normalized chloramine residual. NPNV, pH 8.0, 20°C, 4 mg/L chloramine residual at 24 hrs.**

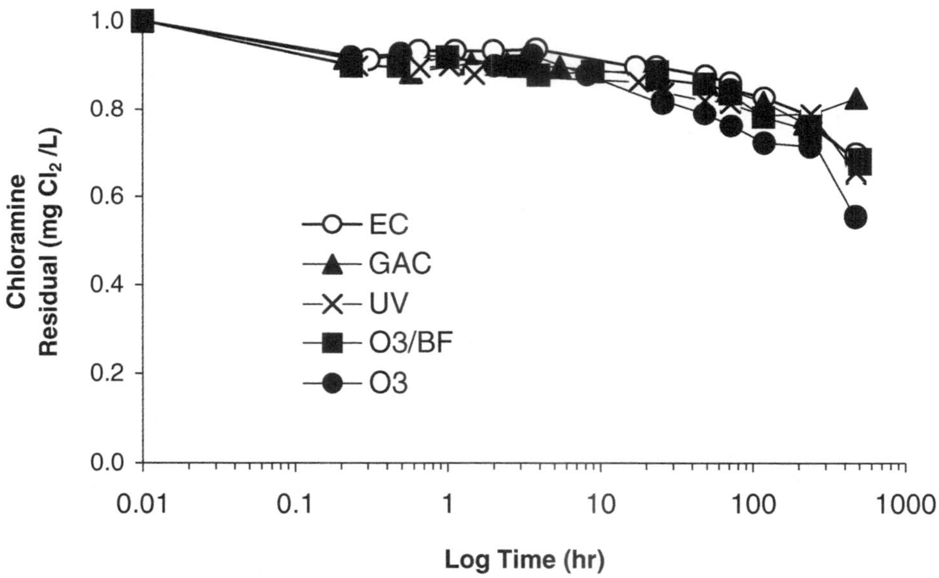


**Figure 3.54 Effect of treatment on normalized chloramine residual. Winchester, pH 8.0, 20°C, 2 mg/L chloramine residual at 24 hrs.**

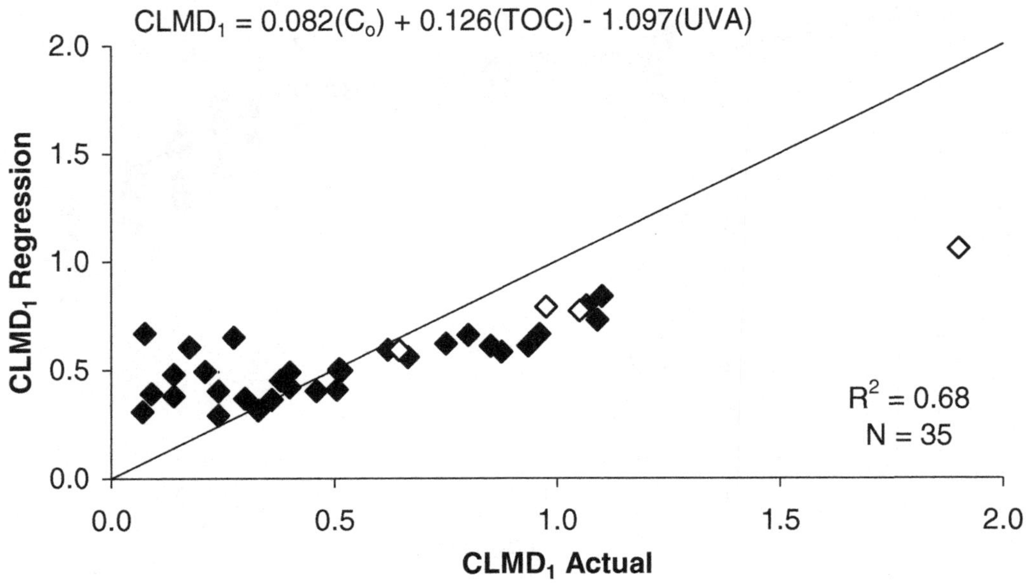


**Figure 3.55 Regression vs. actual 1-day chloramine demands. Solid symbols: values used for regression. Hollow symbols (2 NPNV Raw, 1 NPNV $O_3$, 1 Winchester $O_3$): not used in regression. Betasso, NPNV, Winchester source waters; EC, UV, GAC, $O_3$, $O_3$/BF; pH 8.0, 20°C.**

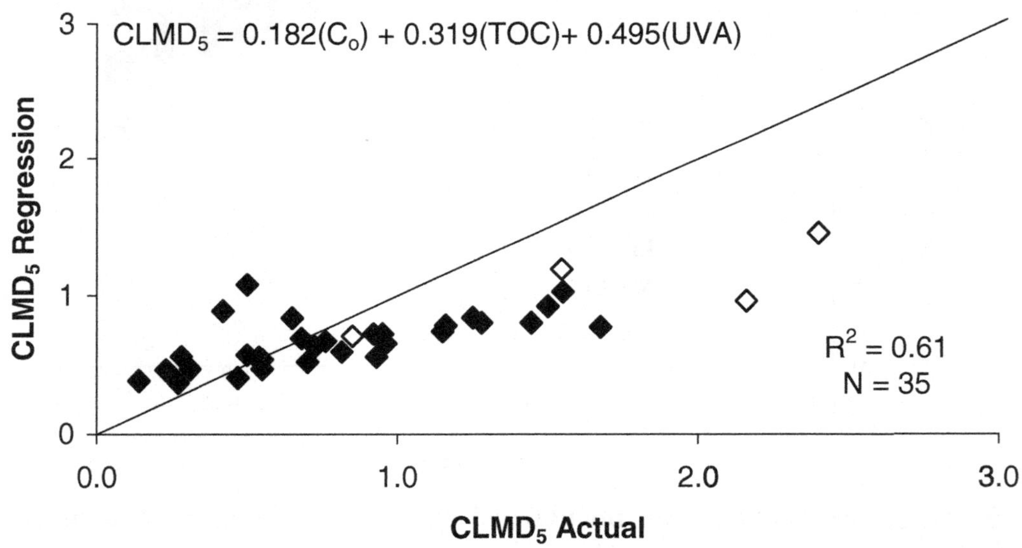


**Figure 3.56 Regression vs. actual 5-day chloramine demands. Solid symbols: values used for regression. Hollow symbols (2 NPNV Raw, 1 NPNV $O_3$, 1 Winchester $O_3$): not used in regression. Betasso, NPNV, Winchester source waters; EC, UV, GAC, $O_3$, $O_3$/BF; pH 8.0, 20°C.**

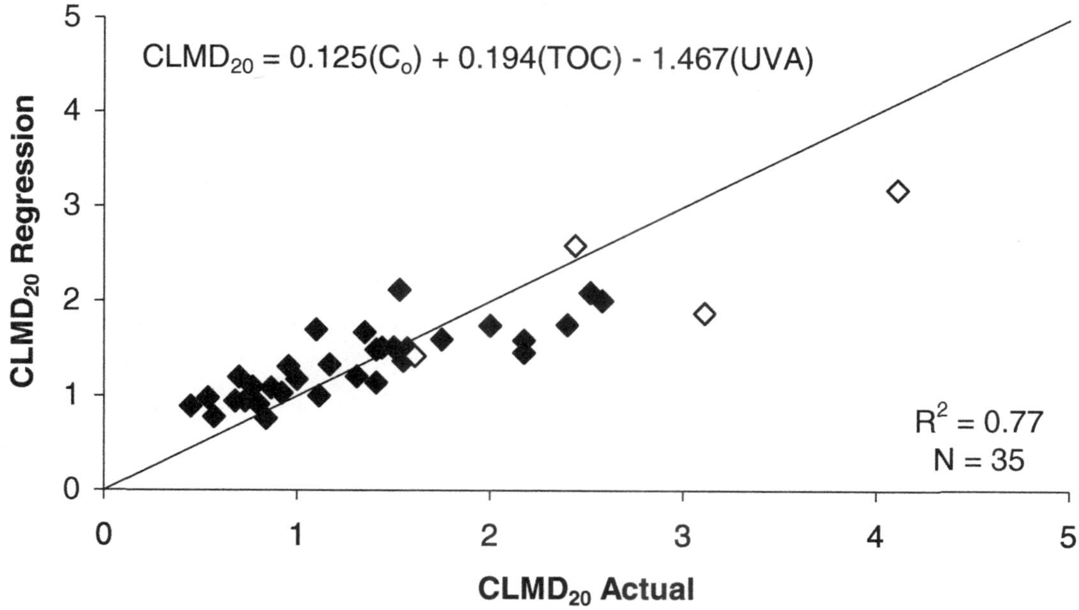


**Figure 3.57 Regression vs. actual 20-day chloramine demands. Solid symbols: values used for regression. Hollow symbols (2 NPNV Raw, 1 NPNV $O_3$, 1 Winchester $O_3$): not used in regression. Betasso, NPNV, Winchester source waters; EC, UV, GAC, $O_3$, $O_3$/BF; pH 8.0, 20°C.**

# CHAPTER 4
# PILOT-SCALE TESTING

## OBJECTIVE

The objective of the work described in this Chapter was to measure the rate of reaction of chlorine in a simulated pipe environment for water receiving different advanced treatments. The simulated pipe environment was designed to represent flow conditions within a ductile iron pipe that had been subject to significant corrosion and biofilm buildup. The following key questions were examined:

1. Does the type of advanced treatment have any effect on chlorine reaction kinetics for water flowing in an iron pipe?
2. What are the relative rates of bulk demand and wall demand for both free chlorine and chloramine?
3. How do the kinetics for free chlorine differ from those of chloramine?

## METHODS AND MATERIALS

### Distribution System Simulator

The experiments were performed at the USEPA's Test and Evaluation (T&E) Facility in Cincinnati, Ohio, using their Distribution System Simulator shown in Figure 4.1. The particular pipe used in this study was an 88-foot (27-meter) long loop of 6-inch (150-mm) diameter, unlined ductile iron pipe equipped with a recirculation pump and a heat exchanger cooling system (Figure 4.2). The pipe has been in service for several years and shows evidence of considerable uniform corrosion and tubercle buildup on its inside surface (Figure 4.3). Examination of coupons placed flush with the pipe-wall also confirmed the presence of a significant amount of biofilm (Meckes et al. 1999). A 1,000-gallon feed tank mounted 20 feet above the pipe loop was used to feed test water to the simulator by gravity. When kinetic experiments are not being performed, the loop is kept in maintenance mode by continuously feeding tap water to it at 1 gpm while maintaining an 88 gpm recirculation rate in the loop. This results in a 1-day residence time for water in the loop and with a free chlorine residual of approximately 0.2 mg/L.

### Experimental Procedure

Each experiment began by adding chlorine (and aqueous ammonium chloride solution for chloramine tests) to a volume of treated test water stored in the 1,000-gallon feed tank. The chlorinated water was held in the feed tank for a 2-hours to simulate the typical contact time achieved in a treatment system's clearwell or contact chamber prior to entry into the distribution system. The exception to this was for ozone treated water, where no holding time after chlorination was used. The test water was then introduced into the pipe loop by simultaneously opening the feed line from the tank and the loop's drain lines. This permitted the new test water to gradually displace the old water from the pipe without introducing any air or disturbing any sediment.

**Table 4.1**
**Pilot-scale analytical methods**

| Analyte | Instrument | Method number |
|---|---|---|
| pH | Orion pH Meter | EPA 150.1 |
| Alkalinity | Metrohm 751 GPD Titrino Autotitrator | SM 2320b |
| Ammonia | Alpkem RFA/2 Autoanalyzer | EPA 350.1 |
| TOC | O-I Analytical 700 TOC Analyzer | SM 5310 C |
| UVA at 254 nm | Carey 50 Spectrophotometer | SM 5910 |
| Total and Free Chlorine | Hach DR-2000 Spectrophotometer | EPA 330.5 |

Tracer tests showed that it took approximately 5 pipe volumes of new water to completely replace the old water in the pipe in a time period of about 20 minutes. After this, the inlet and drain lines were closed and the test water was recirculated under turbulent flow conditions of 1.0 ft/s (0.3 m/s) velocity at a Reynolds Number of 50,000 for a period of 8 hours. The test volume of water (approximately 130 gallons (500 L)) thus made one pass around the loop every 90 seconds and the 8-hour test period was equivalent to sending this volume of water down a 5-mile (8.7 km) stretch of pipe.

Close to the end of the pipe filling period, amber glass bottles were filled headspace-free with test water from the feed tanks. These bottles were stored in a constant temperature bath maintained at the same temperature as the pipe loop water. The temperature of water in the pipe loop was maintained at approximately the same temperature as the original feed water by adjusting the flow rate of the refrigerant used in the cooling system. Samples were withdrawn from the pipe loop for analysis every hour over an eight hour testing period. The bottle samples were analyzed for chlorine and chloramine residual periodically over a one to two week period. Temperature, pH, and turbidity in the pipe loop were also monitored to check for consistent water quality conditions. At the start of each experiment, the alkalinity, TOC, and UVA of the test water fed into the pipe were also measured.

Table 4.1 lists the methods used for the various chemical analyses. A commercial grade of sodium hypochlorite was used to chlorinate the test waters. Reagent grade ammonium chloride was used to chloraminate the water in selected tests. All glassware was prepared by rinsing several times with deionized (DI) water and heating in a muffle furnace at 550°C for 4 hours. Bottle caps were soaked in a high-strength chlorine solution for 24 hours, rinsed 3 times with DI water, soaked in DI for another 24 hours, then rinsed 6 times in DI water.

### Experimental Conditions

The test water used in this study was partially treated Ohio River water obtained from the Greater Cincinnati Water Works. This water had been settled and filtered but not chlorinated. Six different batches of this water were obtained over the 8-month period during which tests were conducted. Each batch consisted of about 4,000 gallons which was stored in a 5,000-gallon stainless steel tank at the T&E facility. There was essentially no ammonia present in this water so that after chlorination only free chlorine was present.

The following advanced treatments were applied to the test water for the experiments:

- GAC adsorption using an Alamo Water C-161A carbon filter cartridge rated at 6.4 gpm/ft$^2$ and empty bed contact time of 3.3 minutes
- ozonation ($O_3$) using a ClearWater Tech Model P-2000 ozone generator rated at 2.8 grams/hr of $O_3$ at 15 scfh air flow and operated to supply a dose of 4 mg/L of $O_3$ to process water flowing at 3 gpm
- reverse osmosis (RO) treatment using an Osmonics E2-1690 machine with a FASTEK S2540 membrane operating at a feed rate of 2.3 gpm with 50 percent recovery

### Estimation of Kinetic Parameters

The following first-order kinetic parameters were estimated from the time series data collected for each experiment:

- the overall decay constant, $K_{total}$
- the bulk water decay constant, $K_{bulk}$
- the solution wall reaction constant, $K_{wall}$

These coefficients were either for free chlorine or chloramine, depending on the experiment. The overall decay constant was estimated by using linear least squares regression with intercept set equal to 0 of the quantity $\ln(C_t/C_0)$ against time, where $C_t$ is the pipe loop chlorine concentration measured at time t and $C_0$ is the pipe loop chlorine concentration at time 0. The bulk water decay constant was estimated in a similar fashion except using chlorine concentrations measured from the bottled samples. The wall reaction constant was then computed as the difference between the overall and bulk water constants according to Equation 4.1.

$$K_{wall} = K_{total} - K_{bulk} \qquad (4.1)$$

where $K_{wall}$ = pipe-wall solution reaction constant (1/time)
$K_{total}$ = overall reaction constant (1/time)
$K_{bulk}$ = bulk water reaction constant (1/time)

Because all experiments were conducted at the same flow rate using the same pipe, the question of difference in mass-transfer effects between experiments was not a factor. Although efforts were made to conduct all of the experiments at the same temperature and pH, outside factors caused temperatures to range from 21 to 26°C and pH ranged from 7.8 to 8.5. Although these variations could have some effects on the chlorine decay kinetics, they were not systematically accounted for in this work.

## PRELIMINARY EVALUATIONS

### Tracer Test

A tracer test was conducted prior to making any experimental trials to confirm that the displacement method used for filling the pipe loop with test water would be effective. Previous experience had shown that simply draining the loop and then filling it with new water creates air bubbles and stirs up excessive turbidity as the recirculation pump must be operating during the filling process. Sodium chloride was used as the tracer material and conductivity was used to measure its concentration. At the start of the test, the pipe loop contained Cincinnati tap water with conductivity of 433 μS/cm. The 1,000 gallon feed tank was filled with water at conductivity of 849 μS/cm. With the recirculation pump running at 88 gpm, the valves on the feed line from the feed tank and on the pipe loop's drain line were opened simultaneously. The resulting time history of conductivity in the pipe loop is shown in Figure 4.4. The original water in the pipe was essentially fully displaced after introducing 600 gallons (5 pipe volumes) of new water, which took 18 minutes.

### Repeatability Tests

Chlorine demand tests were conducted in the pipe loop under identical conditions to assess repeatability. The test water was Cincinnati tap water that was GAC-treated on site to remove any residual chlorine and adsorbable organics. The GAC treatment reduced the bulk chlorine demand to a negligible amount so that the tests could focus on the variability inherent in the wall reactions. The first pair of tests, spaced two days apart, was conducted at an initial chlorine level of 6 mg/L. A second pair of tests was conducted at an initial chlorine level of 3.1 mg/L, spaced one week apart. Between the tests in the second pair, the loop was kept in maintenance mode. The resulting time history of chlorine concentrations is displayed in Figure 4.5 for both Test 1 (A and B) and Test 2 (A and B). These tests demonstrate that to achieve the best repeatability a sufficient amount of time should be maintained between kinetic tests to allow the pipe surface to recover to its pre-test condition.

## EXPERIMENTAL RESULTS

### Free Chlorine Experiments

A set of 8 experiments was conducted using free chlorine as the disinfectant under four different types of treatment following conventional sedimentation/filtration (no further treatment, ozonation, carbon adsorption, and reverse osmosis) and two initial chlorine levels (nominally 2 and 4 mg/L). The initial chlorine levels are the concentrations of free chlorine sampled from the pipe loop at the start of the experiment, after a 2-hour holding time in the feed tank and approximately 20 minutes of fill time.

Table 4.2 summarizes the conditions and kinetic results for each experiment. Figure 4.6 plots the chlorine decay measured in the pipe loop over time for the high chlorine tests while Figure 4.7 does the same for the low chlorine tests. The chlorine decay in the bottle tests over time are shown in Figures 4.8 and 4.9 for the high and low chlorine tests, respectively. Figure 4.10 displays a graphical summary of the estimated wall reaction constants by treatment type and initial chlorine level. Figure 4.11 does the same for the bulk water reaction constants. Note there is no "low dose" for the bulk water constants for conventional treatment in Figure 4.11. This is because there was a sampling miscommunication, and there was not time to repeat the test to collect the additional sample.

An analysis of variance (ANOVA) statistical procedure was applied to the results shown in Figure 4.10 to see if either the chlorine level or treatment type had any statistically significant

**Table 4.2**
**Experimental conditions and kinetic results for free chlorine decay experiments**

| Treatment/ initial $Cl_2$ | TOC (mg/L) | UVA (1/cm) | ALK (mg/L) | Temp. (°C) | pH | $Cl_2$ (mg/L) | $K_{total}$ (1/day) | $K_{bulk}$ (1/day) | $K_{wall}$ (1/day) |
|---|---|---|---|---|---|---|---|---|---|
| CONV/High | 1.54 | 0.034 | 69 | 25.6 | 7.9 | 6.8 | 1.62 | 0.094 | 1.53 |
| CONV/Low | 1.56 | 0.030 | 66 | 26.4 | 7.8 | 3.2 | 2.38 | N/A | 2.29 |
| $O_3$/High | 1.60 | 0.013 | 57 | 21.6 | 8.0 | 3.8 | 3.00 | 0.048 | 2.95 |
| $O_3$/Low | 1.51 | 0.017 | N/A | 21.4 | 7.8 | 1.9 | 4.71 | 0.109 | 4.60 |
| GAC/High | 0.31 | 0.003 | 61 | 21.6 | 8.0 | 4.2 | 3.00 | 0.005 | 3.00 |
| GAC/Low | 0.72 | 0.006 | 67 | 20.9 | 8.2 | 1.9 | 3.56 | 0.019 | 3.54 |
| RO/High | 0.03 | 0.005 | 5 | 20.8 | 7.8 | 3.9 | 1.07 | 0.007 | 1.06 |
| RO/Low | 0.04 | 0.002 | 3 | 21.3 | 8.5 | 2.1 | 2.75 | 0.005 | 2.75 |

NOTE: ALK = total alkalinity in mg/L as $CaCO_3$ and $Cl_2$ = initial free chlorine in pipe loop at start of experiment

**Table 4.3**
**ANOVA for free chlorine wall reaction rate constant**

| Source of variation | SS | df | SS/df | F | P-value | F crit |
|---|---|---|---|---|---|---|
| Initial Chlorine | 2.691 | 1 | 2.691 | 15.156 | 0.030 | 10.128 |
| Treatment | 5.471 | 3 | 1.823 | 10.271 | 0.044 | 9.277 |
| Error | 0.532 | 3 | 0.177 | | | |
| Total | 8.695 | 7 | | | | |

effect on the wall reaction rate constant for free chlorine. The results are shown in Table 4.3. Both factors are significant at the 5 percent level, although the treatment effect is just barely so.

### Chloramine Experiments

A second series of experiments was performed using chloramine disinfectant instead of free chlorine. The chloramine was generated by adding aqueous ammonium chloride to chlorinated test water in a ratio of 4:1 chlorine to ammonia-N by weight so as to produce essentially all monochloramine. The chloramine concentrations used in the tests were similar to those used in the free chlorine experiments to facilitate comparisons. Actual concentrations of chloramines fed to distribution systems tend to be higher than those of free chlorine because of the lower disinfecting strength of chloramines. Kinetic tests were made using three different levels of treatment (conventional, conventional + ozone, and conventional + GAC) and two different levels of initial chloramine (nominally 2 and 4 mg/L). Reverse osmosis treatment was not used in this phase of the study.

**Table 4.4**
**Experimental conditions and kinetic results for chloramine decay experiments**

| Treatment/ initial $Cl_2$ | TOC (mg/L) | UVA (1/cm) | ALK (mg/L) | Temp. (°C) | pH | $Cl_2$ (mg/L) | $K_{total}$ (1/day) | $K_{bulk}$ (1/day) | $K_{wall}$ (1/day) |
|---|---|---|---|---|---|---|---|---|---|
| CONV/High | 1.57 | 0.048 | N/A | 21.4 | 8.1 | 4.2 | 1.17 | 0.015 | 1.16 |
| CONV/Low | 2.25 | 0.036 | 63 | 21.6 | 8.0 | 2.6 | 1.42 | 0.017 | 1.41 |
| $O_3$/High | 1.53 | 0.027 | 66 | 21.6 | 8.0 | 4.2 | 1.23 | 0.007 | 1.22 |
| $O_3$/Low | 0.75 | 0.016 | 65 | 21.7 | 8.2 | 2.5 | 1.43 | 0.005 | 1.43 |
| GAC/High | 1.72 | 0.031 | 45 | 22.1 | 7.9 | 4.1 | 1.34 | 0.010 | 1.33 |
| GAC/Low | 1.80 | 0.019 | 44 | 22.1 | 7.9 | 1.8 | 1.54 | 0.017 | 1.51 |

NOTE: ALK = total alkalinity in mg/L as $CaCO_3$ and $Cl_2$ = initial free chlorine in pipe loop at start of experiment

**Table 4.5**
**ANOVA for chloramine wall reaction rate constant**

| Source of variation | SS | df | SS/df | F | P-value | F crit |
|---|---|---|---|---|---|---|
| Initial Chloramine | 0.0683 | 1 | 0.0683 | 110.7027 | 0.0089 | 18.5127 |
| Treatment | 0.01923 | 2 | 0.0096 | 15.5945 | 0.0602 | 19.0000 |
| Error | 0.00123 | 2 | 0.0006 | | | |
| Total | 0.08873 | 5 | | | | |

Table 4.4 summarizes the conditions and results for each chloramine experiment. The decay of chloramine in the pipe loop is charted in Figures 4.12 and 4.13 for the high and low chloramine experiments, respectively. Figures 4.14 and 4.15 show the bulk water decay monitored in the bottle tests for the high and low chloramine experiments, respectively. Figure 4.16 compares wall reaction constants for chloramine decay by both treatment type and initial chloramine level while Figure 4.17 does the same for the bulk reaction constant.

As was done for the free chlorine experimental results, an ANOVA was applied to the results shown in Figure 4.16 to see if either the initial chloramine level or treatment type had any statistically significant effect on the wall reaction rate constant. The results are shown in Table 4.5. Disinfectant concentration continues to have a significant effect at the 5 percent level but treatment type does not.

## DISCUSSION OF RESULTS

### Treatment Characteristics

The characteristics of the water treated by the advanced treatment processes behaved as expected. The average values of selected water quality parameters are compared in Table 4.6.

**Table 4.6**
**Average water quality characteristics of alternative treatments**

| Treatment type | TOC (mg/L) | UVA (1/cm) | Alkalinity (mg/L as $CaCO_3$) | pH |
|---|---|---|---|---|
| Conventional | 1.73 | 0.037 | 66 | 8.0 |
| Ozone | 1.66 | 0.020 | 49 | 7.9 |
| GAC | 0.83 | 0.013 | 65 | 8.1 |
| RO | 0.04 | 0.004 | 4 | 8.2 |

Ozone achieved only a slight reduction in TOC but reduced the UV absorbance by nearly 50 percent, indicating the expected conversion of complex humic substances into simpler compounds with fewer aromatic double bonds. There was also a modest reduction in alkalinity. GAC treatment removed about half (52 percent) of the TOC. The remaining organics had a much lower UV absorbance while alkalinity was unchanged. The RO membrane treatment removed essentially all of the TOC, UVA, and alkalinity in the water.

### Concentration Effect

The experimental results revealed a significant effect of initial concentration of chlorine on the rates of decay, both in the bulk solution and at the wall. This behavior can be attributed to the effect that reactants other than chlorine have on the rate of chlorine decay. As shown in Hua et al. (1999), if one assumes that chlorine ($Cl_2$) reacts with some group of reactants (X) according to Equation 4.2, then a second-order reaction rate expression for chlorine can be written as Equation 4.3 when $X_0 >> \alpha C_0$.

$$Cl_2 + \alpha X \Rightarrow products \quad (4.2)$$

$$\frac{dC}{dt} = -k(X_0 - \alpha C_0)C = -k_1 C \quad (4.3)$$

where
$C$ = chlorine concentration at time t
$C_0$ = chlorine concentration at time 0
$k$ = overall reaction rate constant
$X_0$ = concentration of reactants at time 0
$\alpha$ = stoichiometry constant
$k_1$ = apparent first-order reaction rate constant

Thus the apparent first-order rate constant is actually a function of initial chlorine concentration, and it increases as the initial chlorine value decreases. This is consistent with the behavior observed in this study.

### Treatment Effect—Bulk Reactions

Type of treatment had a clear effect on the bulk water reaction kinetics of free chlorine, with the reaction rate constant increasing in the following order: RO → GAC → ozonation → conventional treatment, with the latter two treatments being almost equivalent. The GAC treated water had a decay rate only about 1/10 of that for both conventionally treated and ozonated water. Similar results held for the chloraminated water, but the bulk decay rate for GAC treated water was only about half that of the conventionally treated and ozonated water. Comparing bulk rates between the free and combined chlorinated waters shows that for conventionally treated and ozonated water, chloramine decays 5 times slower as compared to free chlorine. For GAC treated water, chloramine decayed 2 times more slowly than free chlorine.

### Treatment Effect—Wall Reactions

For free chlorine pipe-wall reactions, the rate constants for conventional and RO-treated waters were comparable in value. The constants for the ozonated and GAC treated water were also comparable, but were about twice as high as the other two treatment types. One reason for the higher rates for ozonated and GAC treated water, at least compared to the conventionally treated water, might be the changes in TOC that occur. It has been reported that NOM in water can sequester iron through complexation reactions (Knocke et al. 1994). This behavior can cause problems in achieving high removal efficiencies in oxidative iron treatment systems. One can postulate that under $O_3$ and GAC treatment, the nature of the NOM was either changed (for $O_3$) or reduced (for GAC) in such a way that less ferrous iron from the pipe-wall was complexed and was, therefore, more readily available to react with chlorine. This explains the higher reaction rate of these two processes as compared to conventional treatment. However, this explanation contradicts the kinetic behavior observed for the RO-treated water, which was similar to that of the conventionally treated water, yet had essentially no TOC remaining in it. A potential explanation for the behavior observed in the RO-treated water is that the reduction in total dissolved solids (inferred from the drop in alkalinity) significantly reduced the water's conductivity and thus its ability to conduct electrons which is a necessary part of the corrosion process. Thus the enhancement of chlorine decay by a reduction in TOC was being offset by the reduction in corrosion rate due to reduced total dissolved solids (TDS).

A similar behavior was not observed for water disinfected with chloramine. In this case the wall rate constants were essentially the same across all treatments at a given initial chlorine level. This might reflect the possibility that reactions involving chloramines follow first-order kinetics more closely, being limited only by the concentration of chloramines. In contrast, the free chlorine reactions are more dependent on the rate at which reductants (e.g., ferrous iron) are made available at the pipe wall, which in turn were affected by the amount of TOC and conductivity remaining after treatment. This can be seen when examining the magnitude of the initial concentration effect described earlier. For chloramines, the effect ranged between a 14- to 22-percent increase in rate constant when initial chlorine concentration is reduced from 4 mg/L to 2 mg/L. For free chlorine the effect ranged from 18 to 159 percent, thus implying a higher dependence of reaction rate on the concentrations of the reacting species.

### Disinfectant Effect

Figure 4.18 compares the ratio of the decay rate constants for chloramine to free chlorine by treatment type for both wall and bulk reactions. Each ratio plotted is the average of the ratios from the high and low concentration tests. For wall reactions, the chloramine reacted only 70 percent as fast as free chlorine in the conventionally treated water and only 40 percent as fast in the ozonated and GAC waters. The latter two treatments show slower chloramine decay relative to conventional treatment because the free chlorine rates were higher for ozone and GAC (as previously discussed) and the effect of treatment on chloramine wall demand was negligible. The reductions in bulk decay constants of chloramine relative to free chlorine show more variability than the wall constants. However, given the extremely low values obtained for GAC treated water and having only a single test result available for the conventionally treated water makes it difficult to draw comparisons from these data.

## OBSERVATIONS FROM PILOT-SCALE TESTING

Fourteen kinetic experiments were conducted to measure the effect of advanced treatment on the rate of chlorine decay in a simulated water distribution system consisting of 6-inch diameter unlined ductile iron pipe. Each experiment collected chlorine decay data from a different test water recirculated over an 8-hour period in the simulator. The test waters were also subjected to a longer term bottle test to determine bulk water decay rates. First-order rate constants were estimated for both bulk water and pipe-wall reactions. These were used to compare reaction rates by treatment type, disinfectant type, and disinfectant dose. The following findings and conclusions can be drawn from the results of these experiments:

- Regardless of treatment received or type of disinfectant used, first-order rate constants for both the bulk water and pipe-wall reactions are higher at lower initial chlorine concentrations. This suggests that chlorine decay is not a pure first-order process but is also influenced by the concentrations of the species reacting with chlorine.
- Type of treatment had a statistically significant effect on the rate of pipe-wall demand for free chlorine but not for chloramines. The free chlorine wall rates for ozonated and GAC treated water were about twice those of conventional or RO-treated water. This behavior is believed to be due to the effect that changes in the organic content of water have on its ability to complex iron and the effect that changes in water conductivity have on pipe-wall corrosion.
- For the unlined, ductile iron pipe used in these experiments, chlorine decay was dominated by wall reactions. Rate constants for the wall reactions were 1 to 2 orders of magnitude higher than for bulk reactions for both chlorine and chloramine.
- Wall reaction rate constants for free chlorine ranged from 1.1 to 4.6 $days^{-1}$, depending on chlorine level and treatment. These numbers are consistent with the rates measured by others. The wall rate constants for chloramines ranged only between 1.2 and 1.5 $days^{-1}$. Paired comparisons of wall rates for similar treatments showed the chloramine constants to be only 0.40 to 0.70 times as high as those for free chlorine. Bulk rate constants were also lower for chloramine when compared to free chlorine.

The most significant finding of this work is that free chlorine appears to be more susceptible than chloramine to changes in pipe-wall reaction rates due to changes in the amount and nature of the TOC remaining after treatment. One should keep in mind that these results were obtained for a single metallic pipe that has undergone appreciable corrosion, tubercule buildup, and biofilm colonization. These results are therefore only indicative, and not generally conclusive, of what might be expected if treatment changes are made in systems containing pipes of a similar nature.

**Figure 4.1 Photograph of USEPA distribution system simulator**

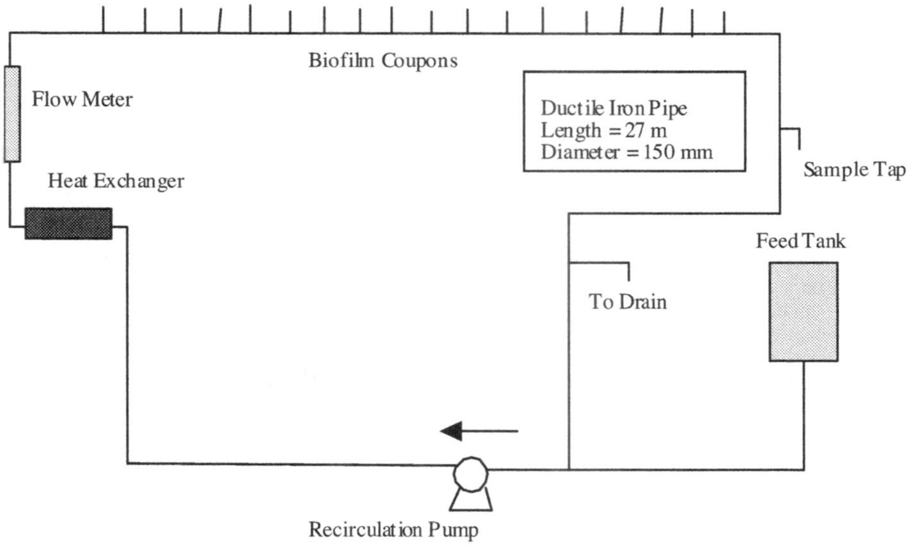


**Figure 4.2 Schematic of distribution system simulator**

**Figure 4.3 Interior surface view of a pipe in the distribution system simulator**

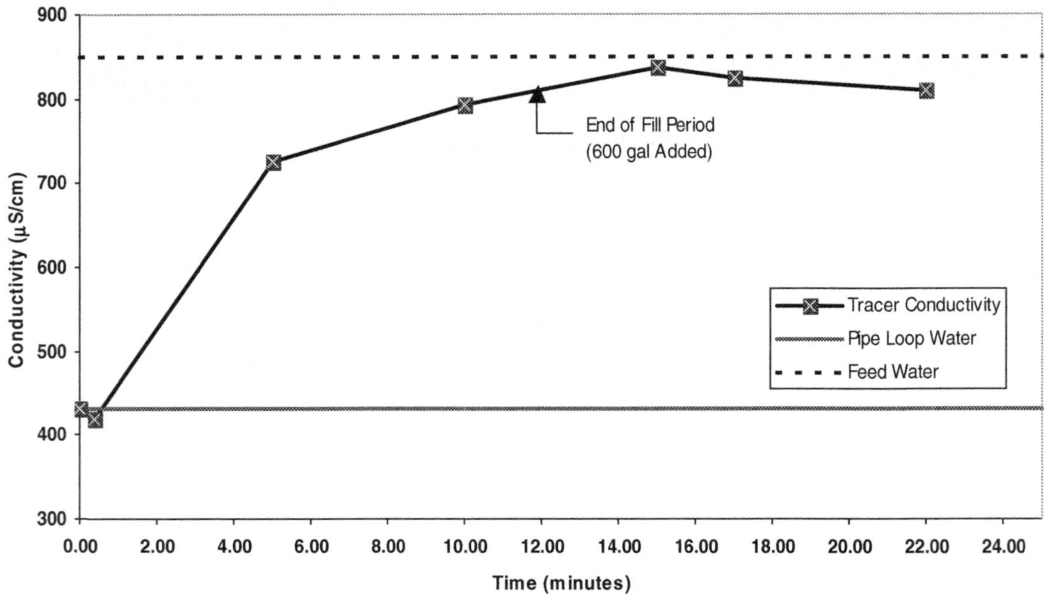


**Figure 4.4 Tracer test results for filling the pipe loop with new water**

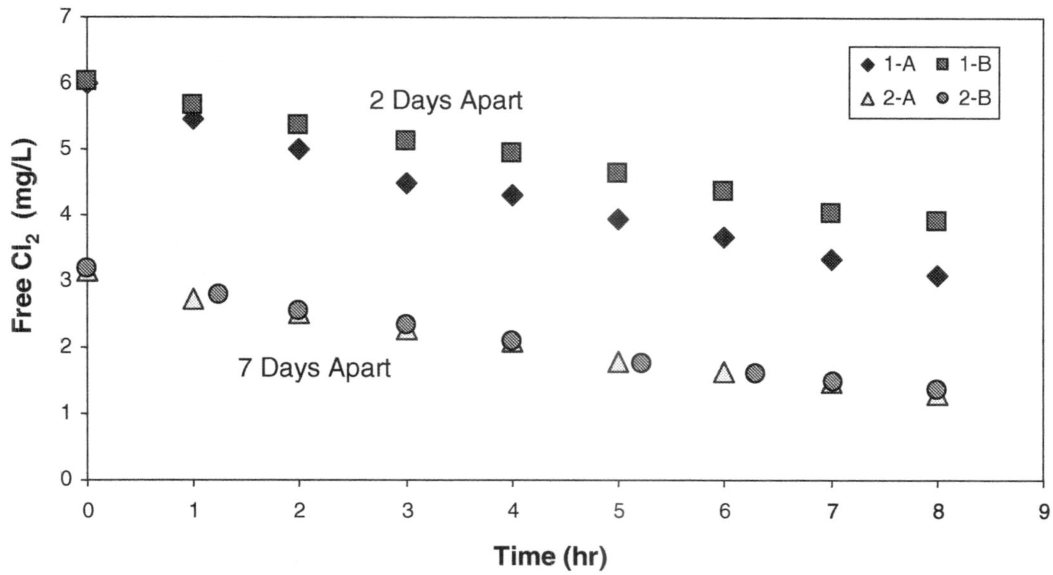


**Figure 4.5 Repeatability of kinetic tests in the pipe loop**

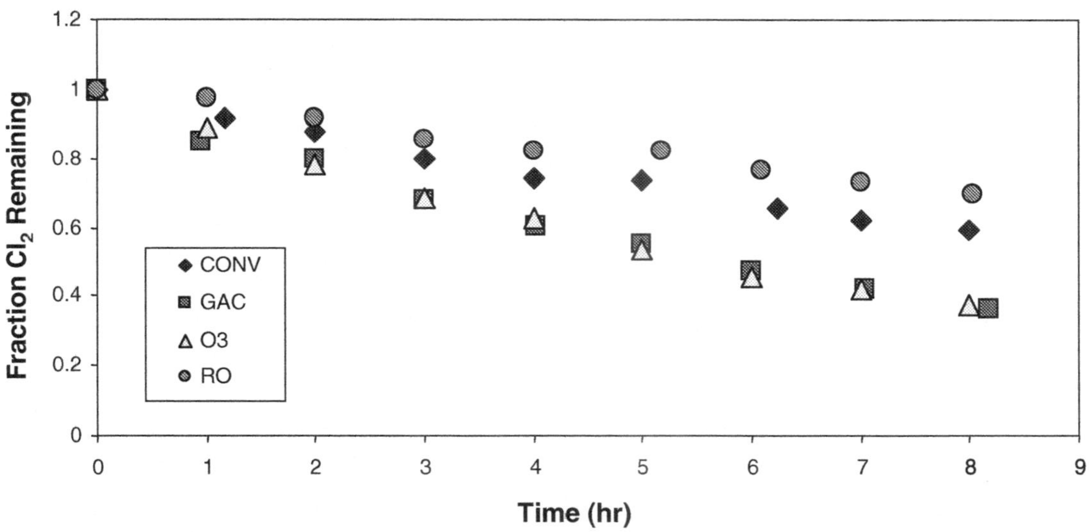


**Figure 4.6 Pipe loop results for high free chlorine experiments ($Cl_2$ = 4 mg/L)**

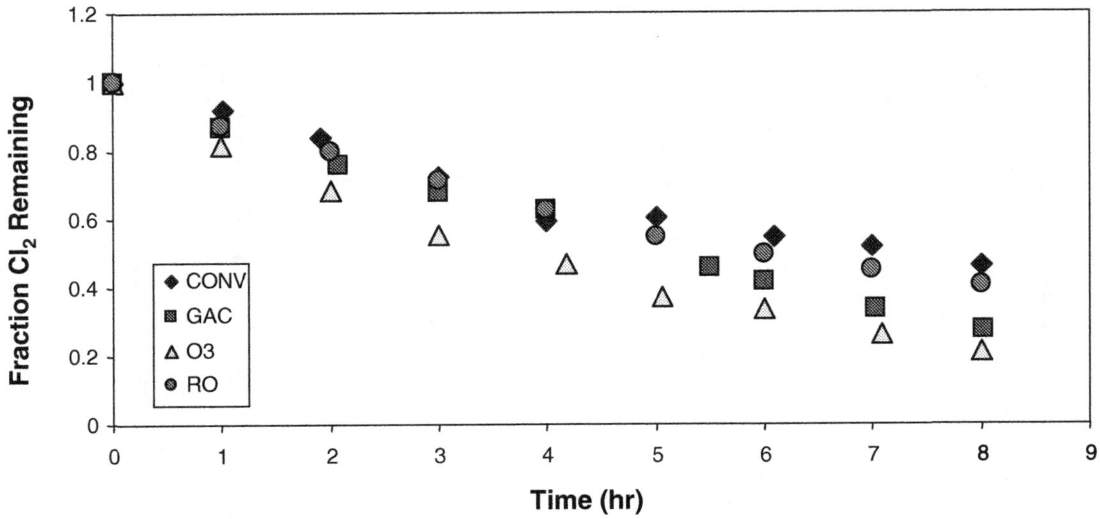


**Figure 4.7 Pipe loop results for low free chlorine experiments ($Cl_2$ = 2 mg/L)**

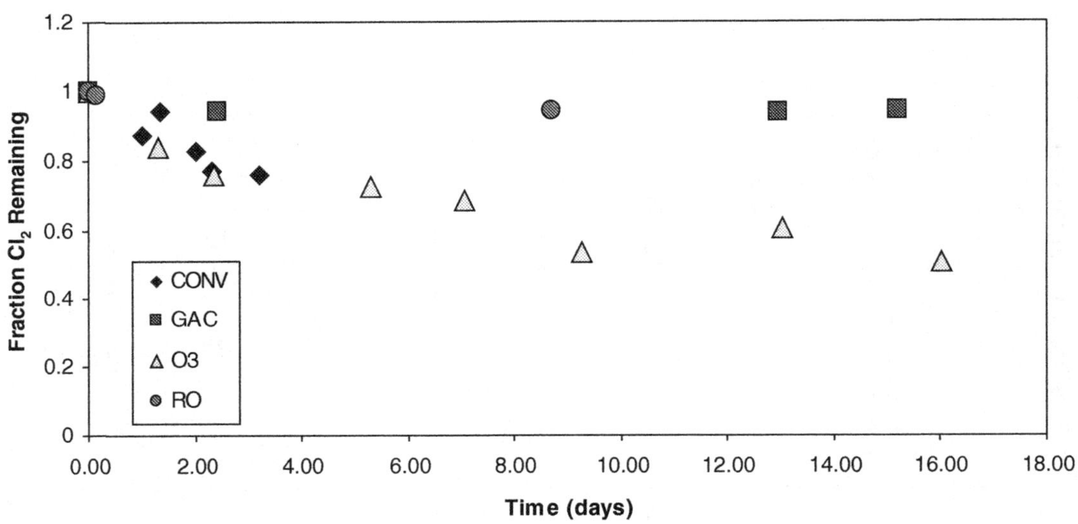


**Figure 4.8 Bottle test results for high free chlorine experiments ($Cl_2$ = 4 mg/L)**

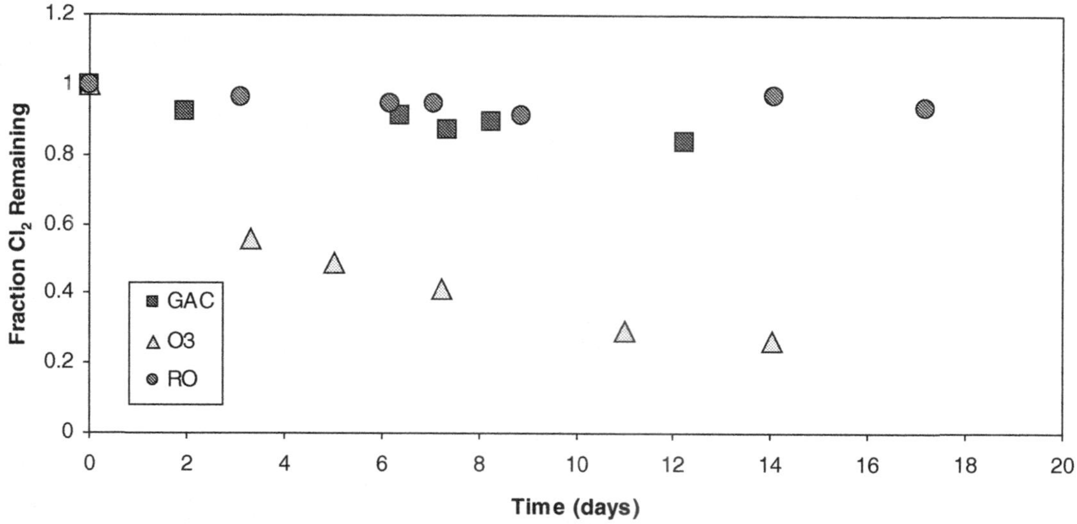


**Figure 4.9 Bottle test results for low free chlorine experiments ($Cl_2$ = 2 mg/L)**

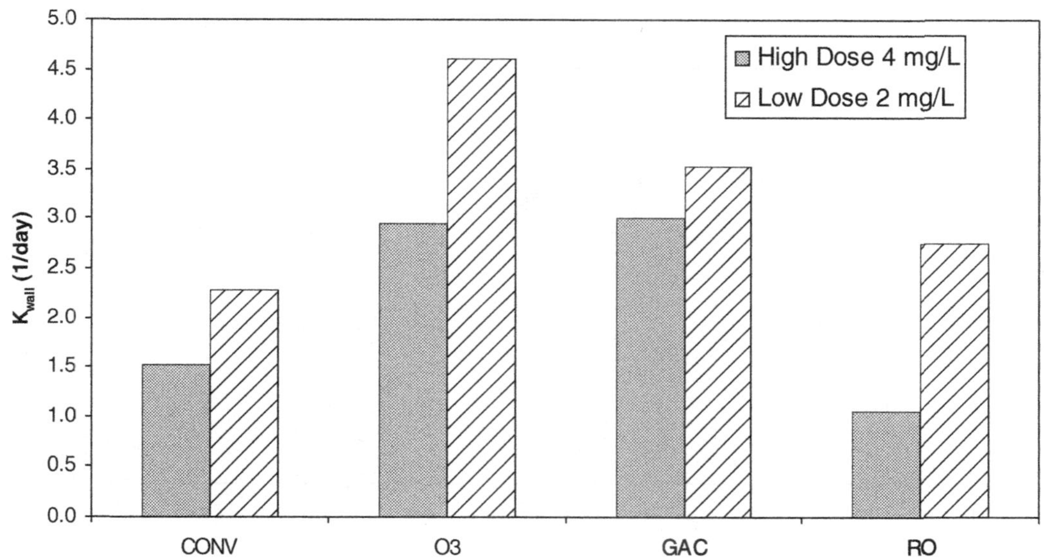


**Figure 4.10 Comparison of pipe-wall reaction constants for free chlorine**

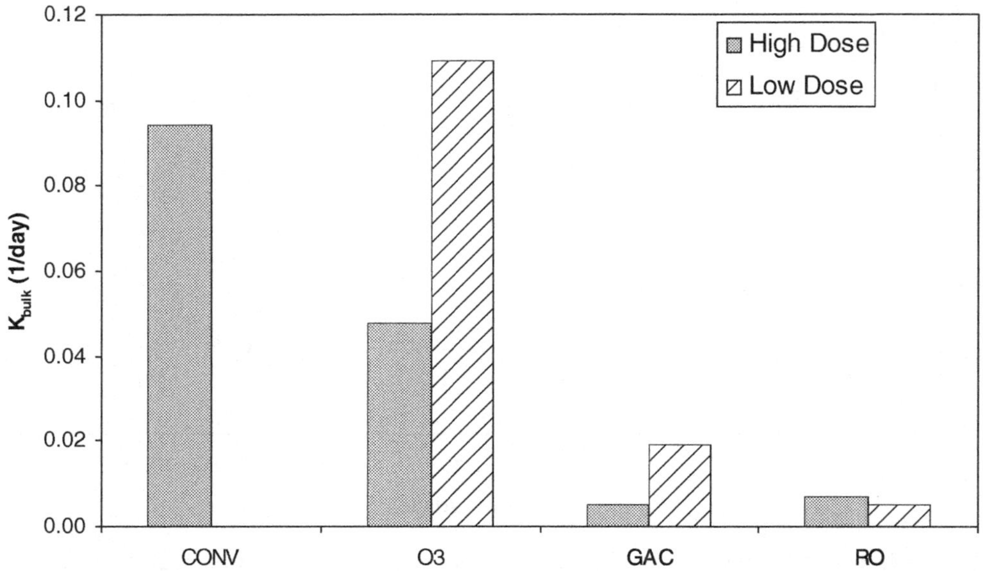


**Figure 4.11 Comparison of bulk water reaction constants for free chlorine**

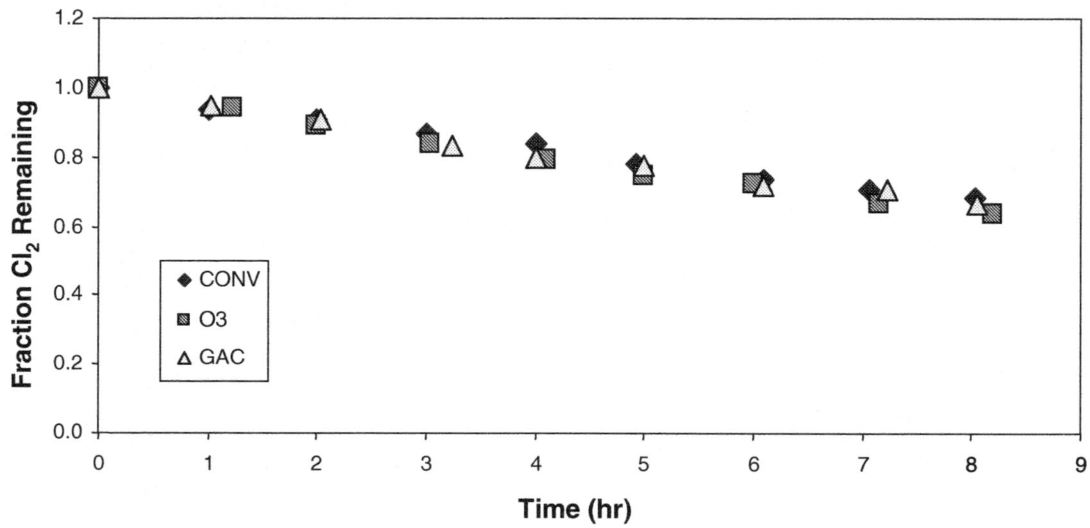


**Figure 4.12 Pipe loop results for high chloramine experiments (4 mg/L)**

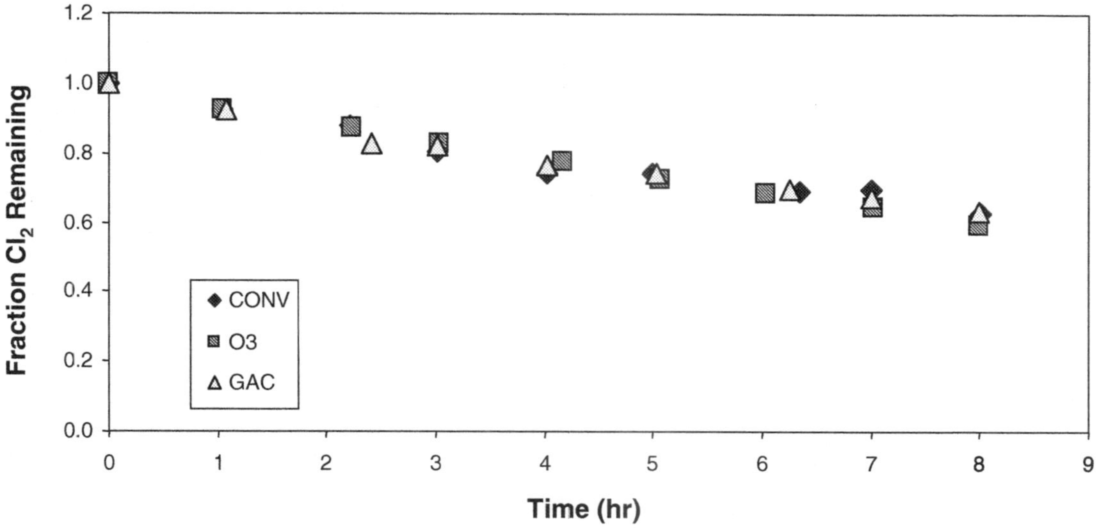


**Figure 4.13 Pipe loop results for low chloramine experiments (2 mg/L)**

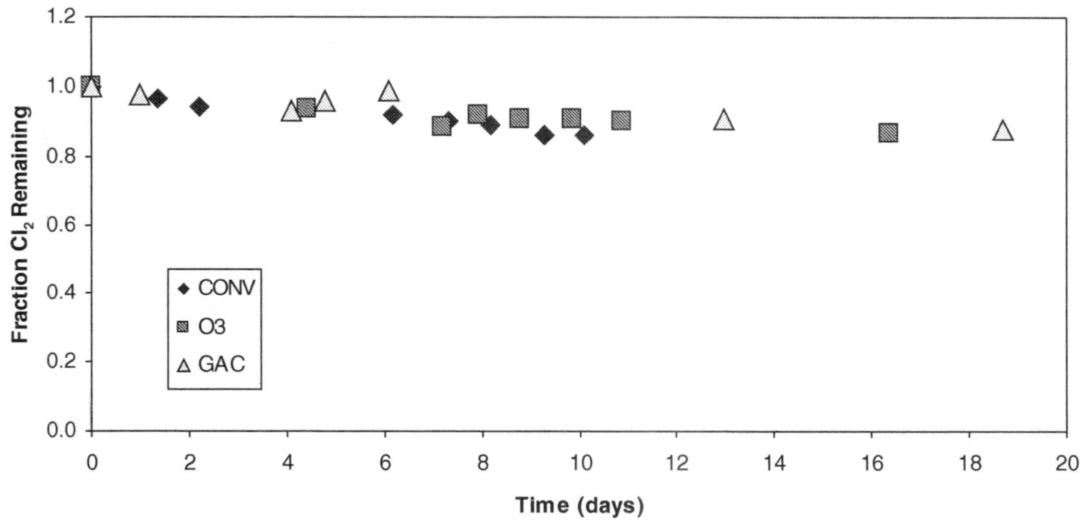


**Figure 4.14 Bottle test results for high chloramine experiments (4 mg/L)**

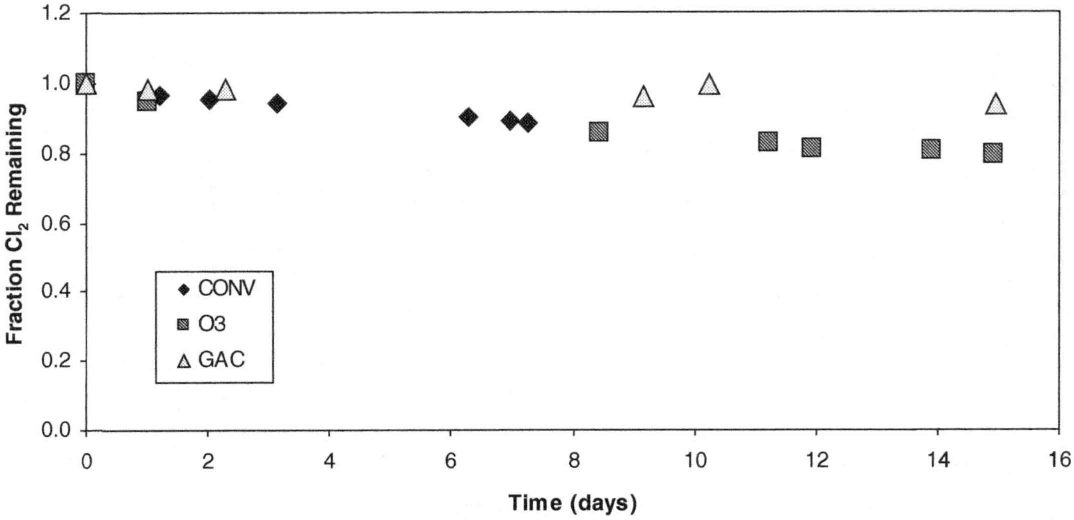


**Figure 4.15 Bottle test results for low chloramine experiments (2 mg/L)**

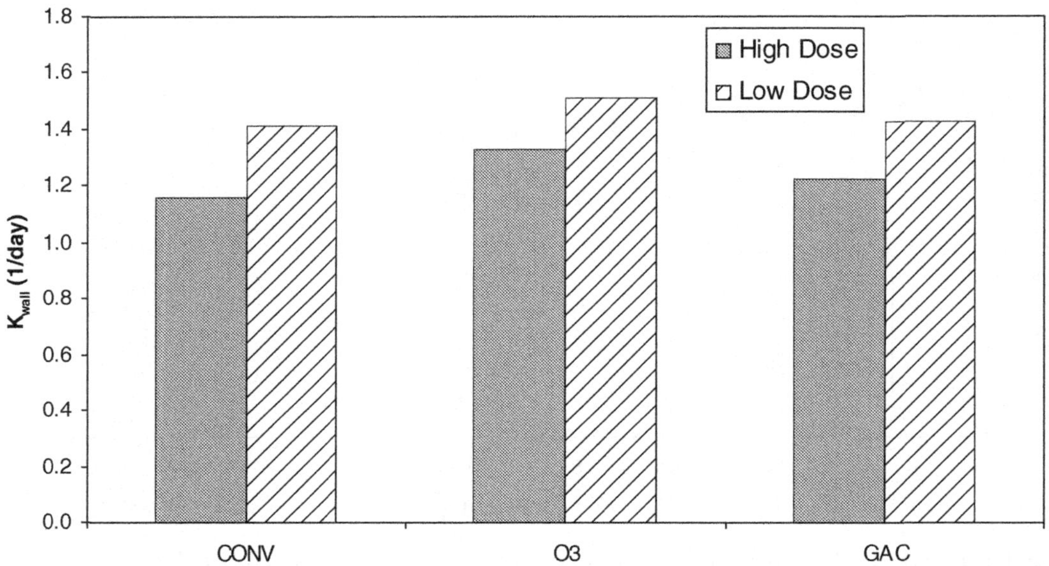


**Figure 4.16 Comparison of pipe-wall reaction constants for chloramine**

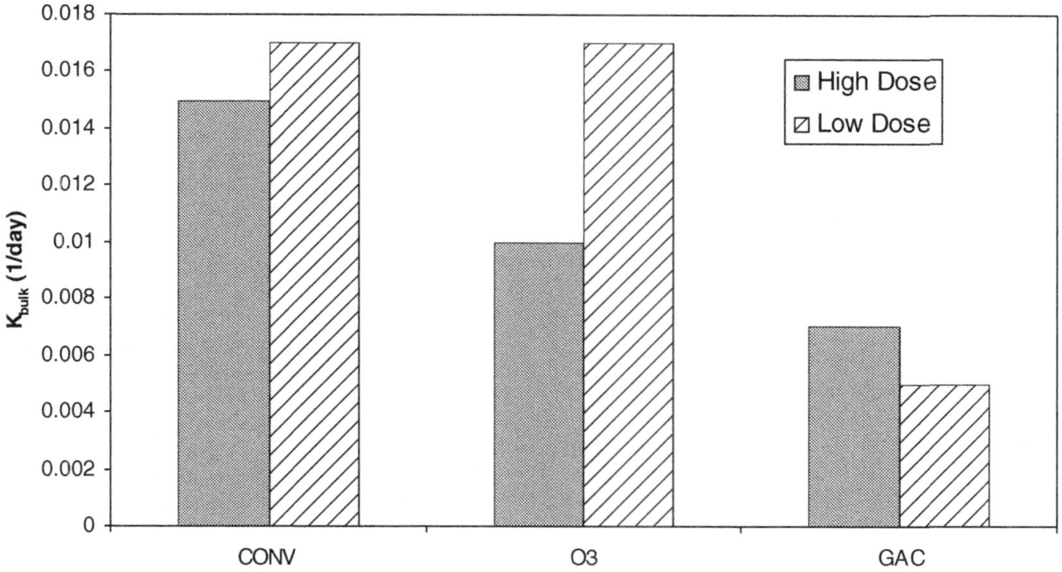


**Figure 4.17 Comparison of bulk water reaction constants for chloramine**

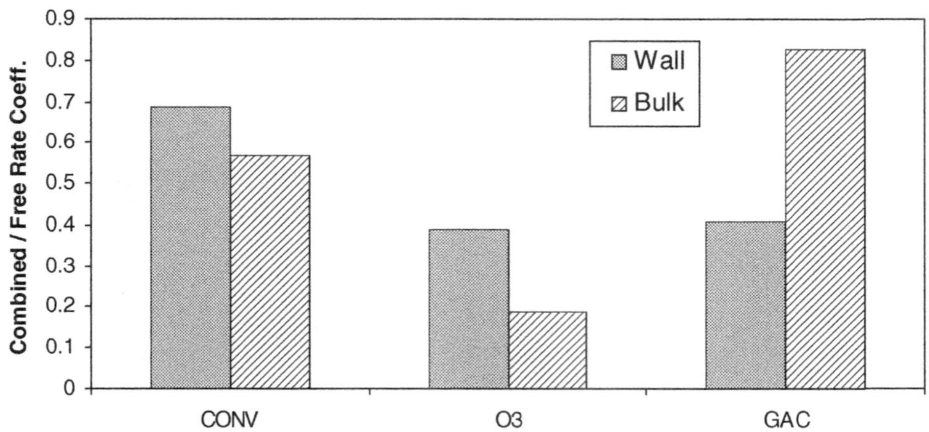


**Figure 4.18 Ratio of chloramine to chlorine rate constants**

# CHAPTER 5
# HYDRAULIC AND WATER QUALITY MODELING

All current distribution system water quality models are limited to track the dynamics of a single species as it is transported and changed via chemical reactions throughout the network of pipes and storage tanks. This is a potentially serious limitation, because all water quality indices result from reactions between chemical species. A single species model implies that distribution system water quality cannot be modeled in a way that mimics the underlying mechanisms.

For example, free chlorine is lost in bulk solution due to oxidation-reduction reactions involving HOCl and $OCl^-$ and NOM. The NOM itself is a heterogeneous mixture of organic compounds (humic and fulvic acids) with varying sizes and chemical characteristics. Current single-species models, however, must model free chlorine loss under the assumption that all other reactants are in excess, and thus their concentrations can be considered to be constant. This limitation is responsible for the widespread observation that decay coefficients of the common single-species, first-order model change significantly as a function of chlorine dose, which is a clear indication of model error. As another example, the mono-, di-, and tri-chloramines result from interactions between free chlorine species and ammonia. These chloramines also interact with NOM, and ammonia is produced from auto-decomposition of chloramines, a fact of significant practical importance in regard to nitrification episodes in distribution systems and storage tanks. Obviously these chloramine reactions involve multiple interacting chemical species, yet current water distribution system models represent their dynamics exactly as for free chlorine dynamics, but with a much smaller decay rate constant. Finally, for the relatively common situation where more than one water source supplies a distribution system, current models are not able to represent the meaningful differences in source water quality as they relate to water quality evolution in the distribution system. Engineers must try to compensate for this limitation by assigning bulk decay rate coefficients to specific pipes, according to which source supplies them. Such an approach has obvious deficiencies when attempting to model distribution system zones where sources blend together, and these zones are often the focus of water quality issues. In addition to these, many other examples exist where multi-species chemical reactions are important, such as the multi-species, multi-step reactions that form THMs, haloacetic acids, and other chlorination by-products.

This research examined the chlorine decay algorithm currently used in EPANET and updated it to include multi-species modeling. Additionally, an innovative method of identifying hydraulic flow paths (network path analysis), and therefore mixing of different water ages and water qualities, was developed. These tools are a direct modification to the *C* code that implements the *Epanet2 Programmer's Toolkit* (Rossman 1999). This toolkit is a set of numerical algorithms that implement the *Epanet2* hydraulic and water quality network algorithms, and a set of function definitions that allows access to them from programs written in the *C* programming language. The *Epanet2* desktop application with which consulting engineers are most familiar, is built using the *Programmer's Toolkit* (it is the numerical "engine" behind the interface), and it has been used worldwide in a host of applications. Thus, by implementing the multi-species modifications to the toolkit code, the multi-species capability becomes accessible to a similar wide range of applications.

The algorithms developed for this project were calibrated and verified with field-scale testing.

## MODELING MULTI-SPECIES CHLORINE DECAY

The multi-species chlorine decay model accounts for interactions between chlorine and NOM and is shown in Equations 5.1 and 5.2.

$$\frac{dC_A}{dt} = k_A C_A C_B \tag{5.1}$$

$$\frac{dC_B}{dt} = k_B C_A C_B \tag{5.2}$$

where species 'A' represents free chlorine and species 'B' represents, in a phenomenological way, the oxidation state of NOM. This model is not original and, in fact, is a prototypical two-species chemical dynamics model for which an analytical solution is known (Levenspiel 1972). Its use to model chlorine decay kinetics was investigated by Clark (1998).

Figure 5.1 shows laboratory measurements of free chorine concentration representing bulk chlorine decay for treated water from South Norwalk, Connecticut (collected as part of a previous AwwaRF project on booster chlorination). The solid line shows a typical 1st order, single-species model that represents the free chlorine decay quite well from 4 hours after dosing to approximately 3 days after dosing at which time the water was rechlorinated to approximately 3.5 mg/L, simulating the use of booster chlorinators to restore a residual in the distribution system. The same first-order model is applied for the rechlorination period (after 72 hours), and is a poor representation of the experimental observations. The rate of decay is greatly exaggerated, because the original dose has reacted with the NOM, and this has affected the decay rate significantly. In short, free chlorine cannot be considered to be in excess, and thus a single-species model is inadequate; some measure of the state of NOM must be included in order to represent this change in decay rates. This inherent multi-species character of these dynamics is emphasized by the dashed lines showing an approximately identical rate of decay (from experimental results) at two very different chlorine concentrations. Because the same rates of decay are observed at very different states (in terms of free chlorine concentration), there must be at least one "hidden" state in the system that is responsible for this because such behavior is inconsistent with a single-species model. This "hidden" state is precisely the state of the NOM, and the above two-species model attempts to account for it through the additional species 'B.'

Figure 5.2 shows the identical data set along with predictions from the two-species model shown in Equations 5.1 and 5.2. Clearly, the two-species model can adequately represent the change in decay rates as the NOM is oxidized by free chlorine. The two dashed lines illustrate that the two-species model can predict identical decay rates at different free chlorine concentrations, because the state of NOM (represented by the "hidden" species 'B') has changed. While this simple two-species model is a phenomenological description, as opposed to a mechanistic one, it appears to be an improvement over the typical first-order, single-species model.

An argument can be made that the experiment described briefly above represents the typical, as opposed to the unusual, dosing experience of water in distribution networks. Consider the simple network illustrated in Figure 5.3, which consists of a single source on the left, which is looped via a "short" and "long" side to a single outlet on the right. Such looping of water paths is ubiquitous in urban water distribution networks. In this typical case water at the outlet node on the right is a mixture of water with short and long residence times. The water with long residence times will have relatively less chlorine and more oxidized NOM, while that with short residence times will have relatively more chlorine and less oxidized NOM. How are the chlorine dynamics described after the water mixes and leaves via the outlet on the right? In essence, one could describe this mixing as a rechlorination of water traveling the "long" side with free chlorine present in water traveling the "short" side, thus approximating to some degree the above experiment – without booster chlorination.

A series of simulation experiments with very simple distribution networks has explored the differences between network predictions using first-order and two-species chlorine decay models. These simulations were obtained using the general extension to the *Epanet2 Programmer's Toolkit,* and thus, also served to test the algorithm modifications using networks for which the types of results could be predicted.

Figure 5.4 shows a network with a single source, a single consumer demand, and a tank in between (a fairly typical network topology). Depending on the consumer demand rate, the water at the consumer is derived directly from the source or is a blend of water from the source and the storage tank. The embedded figure shows the fraction of water at the consumer node that derives from the tank as a function of time for a one-day period. The average residence times of water from the source to the tank is 1/4 day, from the tank to the consumer node is 2-3/4 days, and within the tank is 4 days. The chlorine decay kinetics are described by either the first-order or the two-species model, using the same parameter values as were used in the above figures for the laboratory decay experiments.

The results in Figure 5.5 compare numerical simulations of chlorine concentration at the consumer node as a function of time. The differences between the first-order and two-species model are small (but significant) for this particular example, and the differences are largest when the consumer node is receiving the highest percentage of water from the storage tank.

Figure 5.6 shows a similar network example, but the system has been modified to booster chlorinate water that is leaving the storage tank. In this case the simulation results in Figure 5.7 show relatively larger discrepancies between the typical first-order and two-species models, and again these discrepancies are aggravated by long residence time water from the storage tank. Given the above laboratory data, the two-species model is a more adequate representation.

## ALGORITHM MODIFICATIONS TO EPANET

To date, a flexible, multi-species, distribution system water quality model has not been available for testing various water quality hypotheses and models. This section describes an extension of the *EPANET Programmer's Toolkit* (Rossman 1999) to include general multiple-species dynamic modeling. The data structures and algorithms within EPANET were changed to accommodate any number of additional species, and the modeler is required to specify the reaction dynamics in the bulk water and the pipe-wall. Its use will be illustrated by example applications that compare multi-species and standard single-species model results.

EPANET is a popular drinking water distribution system (DWDS) model for which source code is available in the public domain. It performs extended period modeling of hydraulic and water quality behavior within DWDS. The *EPANET Programmer's Toolkit* (Rossman 1999) is a dynamic link library (DLL) that provides several functions allowing users to use *EPANET's* hydraulic and water quality subroutines for their own specific modeling needs.

The multi species modifications are made directly to the *C* language code that implements the *Epanet2 Programmer's Toolkit*. Because both the *EPANET* desktop application and Programmer's Toolkit are widely used among researchers and engineers, the multi-species capability becomes accessible to a similar wide range of applications by implementing the modifications to the *Toolkit* code. The extensions described here allow for propagation and reaction of multi-species within the bulk water, including their reactions with each other and with pipe-wall materials. The modifications do not, however, allow for the dynamics of species that are not dissolved but rather are associated with the pipe surface (e.g., biofilms). The numerical algorithms are standard ones for non-stiff ordinary differential equations (ODEs), and thus are not appropriate for modeling multi-species dynamics where the species are changing on very different time scales (i.e., stiff ODEs).

## Data Structure Modifications

Because the data structures in *EPANET* are designed for the modeling of single-species water quality, they need to be modified to accommodate the tracking of a vector of chemical species of arbitrary size throughout the network, with the number of species specified via an extension to the standard *EPANET2* input data file. The modifications of the data structures are listed below:

1. **Introduction of NumSpecies**. A new global variable NumSpecies is added to represent the number of species in the reaction model. It is defined in header file \vars.h" and has default value 1.
2. **Modification of Array C and R**. Global variables C (node actual quality) and R (pipe reaction rate) are expanded from one dimensional to two dimensional arrays of floats. Both C and R are defined in header file \vars.h". C stores the node quality and gets updated every water quality time step, and R stores the normalized volume-weighted reaction rate (mass/L/day) in each pipe. The first dimension of the modified C and R arrays is the node or link index and the second is the species index, both starting from 1. All modifications of *EPANET Toolkit* Source code for multi-species modeling, the species index always starts from 1.
3. **Modification of Wbulk, Wwall, Wsource and Wtank**. Global variables Wbulk (average bulk reaction rate), Wwall (average wall reaction rate), Wsource (average mass flow) and Wtank (average tank reaction rate) are used to save the average network-wide reaction rates in mass/hour, which are written to the *EPANET* output file at the end of the simulation. They are expanded from type float value to float pointer that addresses an array of dimension NumSpecies.
4. **Modification of MassIn**. The variable MassIn is global in the water quality modeling module (quality.c) and accumulates the mass flowing into the nodes from adjacent segments, which is used in the update of node water quality. It is expanded from one dimensional to two dimensional array of float values. The first dimension of the modified MassIn is the node index and the second is the species index, both starting from 1.
5. **Modification of Ssource**. The structure Ssource, defined in the header file \types.h", is a C language structure to represent water quality source objects. It has four components: base source strength (concentration or mass) C0, pattern index Pat, actual mass flow rate Smass, and source type Type. In multi-species water quality modeling, each

of the species can be added at each source. Therefore, the four components in the structure Ssource are expanded to arrays of size NumSpecies+1 with species index starting from 1.

6. **Modification of Snode**. The structure Snode, defined in the header file \types.h", is a C language structure to represent node objects. The component C0 represents the initial water quality at the node and is expanded to a float array of size NumSpecies+1. This structure also includes a source object with structure Ssource (see above).
7. **Modification of Stank**. The structure Stank, defined in the header file \types.h", is a C language structure to represent tank objects. The component C stores the concentration of the tank and is changed to an array of floats with size NumSpecies+1.
8. **Modification of Sseg**. The structure Sseg, defined in the header file \types.h", is a C language structure to represent pipe segment objects. The component C represents the uniform concentration within the segment and is changed to an array with size NumSpecies+1.
9. **Introduction of new global array QX**. A general purpose array X (float) is used in both hydraulic and water quality modeling, mainly for swapping data. For multi-species water quality modeling, a two dimensional array is needed for data swapping. If the array X is expanded, many codes involved in only hydraulic modeling would be affected. Therefore, a new array QX is introduced and defined in \vars.h". In water quality modeling, X is replaced by QX. The first dimension is the node or link index and the second is the species index, both starting from 1.
10. **Introduction of local variables to water quality modeling**. An integer variable (CurrentSpecies), simulation, is local to water quality simulation is introduced with default value 1. All the DLL functions that set or get water quality parameters apply to the species with index CurrentSpecies. This variable can be changed with a new DLL function ENselectspecies that will be discussed in the algorithm part. Many local variables used in the water quality modeling functions (in \quality.c") must be defined as arrays. To reduce memory allocation time, some of these arrays are defined locally global in \quality.c". Such arrays that need to be mentioned are arrays BR and WR, which store the bulk reaction and wall reaction rate, respectively. The reaction rates are defined in the multi-species dynamics functions that will be discussed in the section of Algorithm Modifications. All arrays are allocated once at the start of water quality modeling. Most of these arrays are one dimensional with size NumSpecies+1, and used for swapping data, such as segment concentration, node and tank concentration and inflow mass or concentration to the nodes.

### Algorithm

The flow of the original *Epanet2* water quality algorithms has been retained, with the essential change being that chemical dynamics within any one network water quality segment are now defined by a user-specified general system of ordinary differential equations:

$$\frac{dC_i}{dt} = f_i(C);\ C_i(t = 0) = C_{i0} \tag{5.3}$$

where $C_i$ = the concentration of the *i*th chemical species
$f_i$ = the function defining the kinetics of the *i*th species as a function of its concentration and that of the other species with which it interacts

This general approach is the standard language for modeling complex chemical kinetics. The code implementing *Epanet2* water quality sources has been modified to accommodate the addition of multiple species to the network, and the toolkit functional interface definition has been extended to access the results of the multi-species dynamics. A standard of backward compatibility has been adopted, such that the multi-species toolkit code will compile and work with any previous application that used the single-species form. Thus, the multi-species capabilities are accessed via newly defined functional interfaces, or via new functional interfaces that modify the behavior of the existing interface functions. An example of the latter is a function that defines which species is returned via existing functions that access nodal concentration simulation results.

The multi-species dynamics, described by the above system of differential equations, is solved currently by either a simple Euler numerical integration scheme, or by various orders of Runge-Kutta numerical integration formulas (a user choice). While these numerical integration schemes have sufficed for relatively simple test examples, more complex schemes may be required if the dynamics are "stiff" – a name given to systems where the chemical species change on very different time scales (such as where one or more chemical species is defined by relatively "fast" kinetics). These characteristics are relatively common to find in chemical reaction systems. Different numerical integration schemes will be investigated on an as-needed basis throughout the project duration.

## METHODOLOGY FOR THE NETWORK PATH ANALYSIS

### Previous Work

Free chlorine loss in drinking water distribution systems is often influenced by pipe-wall chlorine demand. Model prediction of free chlorine residual thus requires estimation of the associated pipe-wall chlorine demand coefficients, whose values depend on pipe material and condition, and can vary throughout the network. In this section, an efficient algorithm is presented for estimating pipe-wall chlorine demand coefficients, based on identifying network hydraulic flow paths. A previously developed Input-Output (I/O) model identities the flow paths and establishes an analytical relationship between chlorine concentrations at path endpoints. Field measurement of chlorine residuals at these endpoints then allows efficient estimation of wall coefficients for all pipes located on the flow paths. Analytical expressions for the sensitivity of chlorine concentration with respect to pipe-wall chlorine demand coefficients are derived and used in a Gauss-Newton parameter estimation algorithm. A simulation example serves to illustrate the method, while a complete set of field results is presented in the next section.

Recent studies (LeChevallier 1999; Karim et al. 2003) emphasize the need to maintain effective chlorine residual in all parts of the DWDS to protect water consumers from microbiological risks. Although an adequate residual may be present at the point of entry to the distribution system, free chlorine is consumed by reactions with NOM (within the bulk flow) and with pipe-wall materials (so-called "wall demand"). Various water quality models can be used to model chlorine concentration within the DWDS, with the most common assumption being first-order chlorine decay kinetics (Biswas et al. 1993; Rossman et al. 1994; Clark et al. 1994). The bulk decay coefficient is a function of temperature, initial chlorine concentration, and organic content in bulk water (Powell et al. 2000) and can be determined by laboratory bottle decay tests. Wall demand coefficients depend on pipe characteristics (Rossman et al. 1994) such as material, age, and rate of corrosion, which can change from pipe to pipe in the same network. Thus field experiments and data processing methods specifically for estimating wall demand coefficients are rarely reported.

Among the limited published research on wall decay coefficient estimation, one study has reported on least-square estimation for a long pipe based on the chlorine concentration measured at both ends (Maier et al. 2000), and another has reported on trial and error calibration using chlorine data sampled throughout a DWDS (Vasconcelos et al. 1997). Both these approaches have limitations for system-wide estimation of pipe-wall demand coefficients. A trial and error method based on typical field sampling does not explicitly identify the pipes that affect sampled water quality, and instead it adjusts wall demand coefficients of all pipes (or of *ad hoc* pipe groups) to study the model response to parameters. A separate water quality simulation is needed for each combination of possible parameter values. Such an approach can be computationally demanding for a large DWDS with many parameters to estimate, and common techniques for estimating parameter uncertainty are not feasible (e.g., Monte Carlo analysis). The single pipe segment approach eliminates these restrictions because the flow path can be obtained by inspection. The single pipe approach does, however, require that the selected pipe be "long enough" so that chlorine measurement errors are insignificant compared to measured chlorine loss within the pipe. In the published United Kingdom study (Maier et al. 2000), a 1.3-mile-long pipe without interconnections was constructed specifically for wall demand coefficient estimation. Such pipes will be rare within the highly looped sections of a typical DWDS.

As part of this research, an efficient parameter estimation algorithm for wall demand coefficient calibration was developed using the I/O model (Shang et al. 2002). The I/O model identifies the pipes constituting the flow paths between upstream and downstream chlorine sampling locations and targets these pipes specifically for wall demand coefficient estimation. Under the assumption of first-order reaction for chlorine decay, chlorine concentration at a downstream location (output) can be expressed as a linear combination of concentrations at upstream locations (inputs), weighted by water quality "impact coefficients" that reflect water dilution at junctions and chlorine loss due to bulk and wall decay. Because the impact coefficients are explicit functions of pipe-wall demand coefficients, the I/O model establishes an analytical relationship between sampled chlorine concentrations. The sensitivity of modeled chlorine concentration to wall demand coefficients can thus be derived analytically rather than numerically, leading to development of efficient and robust gradient methods for parameter estimation. The following section discusses the chlorine decay model, the I/O water quality model, and the basic concepts of input, output, and network flow path. The parameter estimation algorithm is then presented, including necessary revisions to the I/O model and derivation of the sensitivity of modeled chlorine concentration with respect to wall demand coefficient. Finally, first-order estimates of parameter uncertainty caused by chlorine measurement error are presented, followed by an illustrative simulation example.

## Background

### *Chlorine Decay Model*

A model based on mass transfer (Rossman et al. 1994) is used to predict chlorine decay in DWDS. The model assumes first-order reactions of chlorine both in bulk flow and on the pipe-wall. The wall reaction rate is a function of mass transfer rate of chlorine to pipe-wall and therefore depends on pipe geometry and hydraulic conditions. The overall pipe decay coefficient (K) is a function of the bulk coefficient, wall coefficient, mass transfer coefficient, and hydraulic radius (Equation 5.4)

$$K = K_b + \frac{K_f K_w}{R_H(K_f + K_w)} \tag{5.4}$$

where $K$ = pipe decay coefficient
$K_f$ = flow dependent mass transfer coefficient
$K_b$ = bulk decay coefficient
$K_w$ = wall demand coefficient
$R_H$ = hydraulic radius

In this study, it is assumed that the bulk decay coefficient ($K_b$) is known or determined by bottle test and the measured data are used to calibrate only wall decay rate ($K_w$).

### *Input and Output*

In order to calibrate pipe-wall demand coefficients, chlorine needs to be sampled at different locations that serve as either output or input in this research. Output refers to sampling locations where chlorine measurements are used to compare with water quality modeling results for the purpose of wall demand coefficient estimation. Inputs are upstream of outputs, and chlorine concentrations sampled there are treated as independent variables in the water quality simulation. One kind of input, for example, is a treatment plant where chlorine is injected. Inputs, however, are not limited to water sources. They can be any location that water passes before it reaches an output. For example, in Figure 5.8a, water flows from treatment plant S, passes node I1, and arrives at node O1. Both S and I1 can be treated as inputs corresponding to O1. The output, inputs, and network components connecting them actually constitute a sub network in which inputs serve as both water and chlorine sources.

### *Flow Path in DWDS*

In DWDS, water leaving the treatment plants or other sources travels through different components such as pipes, pumps, and storage tanks before reaching a certain consumer node. Chlorine concentrations at a specific location are affected by the reaction happening in these components that constitute the flow paths between water source and sampling location. Considering the complexity in DWDS topology and operation, flow paths change both spatially (for different sampling locations) and temporarily. In Figure 5.8, water leaves the WTP S and reaches the sampling locations O1 and O2 in Figure 5.8a and 5.8b, respectively. The highlighted pipes are all the pipes through which water may travel before arriving at the sampling locations. It can be seen that although these two sampling locations are physically close to each other, the flow paths connecting them to the WTP are quite different. If two designs are implemented to estimate wall demand coefficients, one design that samples chlorine concentration at S and O1 and the other that samples at S and O2, different pipes are going to be targeted. The flow path concept is applicable to any output and corresponding inputs. Knowing the flow path between output and inputs can help exclude the pipes that do not affect output concentrations and can help construct a much simpler water quality model for output concentration simulation and wall demand coefficient estimation for pipes within the sub network. For a single pipe segment, it is easy to do because the flow path from the input (upstream end) to output (downstream end) is obvious. However, in a DWDS, chlorine concentration can be sampled anywhere, and the identification of flow paths between different sampling locations is not a convenient job without the I/O model.

***I/O Model***

The I/O model developed by the authors (Shang et al. 2002) uses a particle backtracking algorithm (PBA) that transports a water particle reaching an output location upstream and divides it into sub particles at junctions until all particles generated arrive at an input. It is a Lagrangian model that runs in reverse time and can easily identify the pipes constituting the flow paths between inputs and an output. In such an I/O model, the output concentrations can be modeled as a variable, depending on input concentrations, hydraulic conditions in the connecting pipes, and physical characteristics of the connecting pipes. It should be noted that the I/O model, or the sub network constructed by PBA, must be closed. This means water reaching the output must pass one of the inputs. In Figure 5.8b, output O2 and input I2 can not constitute an I/O model because part of the water reaching O2 does not pass the input. Output O2 needs both inputs I2 and I3 to make the I/O model closed.

Under the assumption of chlorine first-order decay, the I/O model describes the following linear relationship between the output and its inputs:

$$c_o(T) = \sum_{j=1}^{N} \gamma_j c_{ij}(T_o - t_j) \qquad (5.5)$$

where
$c_o$ = chlorine output concentration
N = number of flow paths between inputs and output
T = output time
$c_{ij}$ = chlorine input concentration for path j
$t_j$ = water travel time or time delay of path j
$\gamma_j$ = impact coefficient of travel path j

The relationship in Equation 5.5 is exact for first-order kinetic models, as shown by Zierolf et al. (1998) and (via a linear superposition argument) Boccelli et al. (1998). The unitless impact coefficient $\gamma_j$ takes the chlorine decay in pipes and storage tanks and flow mixture at junctions into account and is actually the sensitivity of output concentration to path input concentration. The PBA dynamically calculates the number of paths N and their time delays and impact coefficients in the particle backtracking process.

The formulation of the I/O model is illustrated using a simple network without a tank and under steady hydraulic condition. The network topology is shown in Figure 5.9, and the pipe length l, overall first-order decay coefficient K, flow rate q, velocity v, and water travel time t from upstream node to downstream node are shown in Table 5.1 with arbitrary units. The concentration at node o, a, and b can be expressed in terms of the concentration at directly connected upstream nodes:

$$c_o(T) = c_a(T - t_1) \times \exp(-K_1 t_1) = 0.819 c_a(T-1)$$

$$\begin{aligned} c_a(T) &= \frac{q_2}{(q_2+q_3)} \times c_s(T-4) \times \exp(-K_2 t_2) + \frac{q_3}{(q_2+q_3)} \times c_b(T-4) \times \exp(-K_3 t_3) \\ &= 0.4 \times 0.670 \times c_s(T-4) + 0.6 \times 0.670 \times c_b(T-4) \\ &= 0.268 c_s(T-4) + 0.402 c_b(T-4) \end{aligned}$$

$$\begin{aligned} c_b(T) &= \frac{q_4}{(q_4+q_5)} \times c_s(T-3) \times \exp(-K_4 t_4) + \frac{q_5}{(q_4+q_5)} \times c_s(T-2) \times \exp(-K_5 t_5) \\ &= 0.333 \times 0.406 \times c_s(T-3) + 0.667 \times 0.549 \times c_s(T-2) \\ &= 0.135 c_s(T-3) + 0.366 c_s(T-2) \end{aligned}$$

Treating node b and s as the inputs, there are two flow paths from the inputs to node o: pipe 3+1 and 2+1, and the I/O model can be expressed as: $c_o$(T) = 0.819$c_a$(T-1) = 0.819(0.268$c_s$(T-5) + 0.402$c_b$(T-5)) = 0.219$c_s$(T-5) + 0.329$c_b$(T-5). Because concentration at node b depends linearly on the concentration at node s, $c_o$ can also be expressed as: $c_o$(T) = 0.219$c_s$(T-5) + 0.329$c_b$(T-5) = 0.219$c_s$(T-5) + 0.329(0.135$c_s$(T-8) + 0.366$c_s$(T-7) = 0.219$c_s$(T-5)+0.120$c_s$(T-7)+0.045$c_s$(T-8). It can be seen that there are three paths in this three inputs one output model: pipe 4+3+1, 5+3+1 and 2+1. The water reaching the node o at T is comprised of three parts that leave input s at time T-5, T-7, and T-8 respectively. The contribution of a path's input concentration to the output concentration is impact coefficient defined previously. The three paths in this example have impact coefficients of 0.219, 0.120, and 0.045.

## Parameter Estimation

### *Mathematical Problem Statement and Key Assumptions*

A traditional least squares approach to parameter estimation is taken, with the objective function to be minimized:

**Table 5.1**
**Data for the I/O model illustrative network**

| Pipe | Length | Decay rate | Flow rate | Velocity | Travel time |
|---|---|---|---|---|---|
| 1 | 1 | 0.2 | 5 | 1.0 | 1 |
| 2 | 2 | 0.1 | 2 | 0.5 | 4 |
| 3 | 2 | 0.1 | 3 | 0.5 | 4 |
| 4 | 3 | 0.3 | 1 | 1.0 | 3 |
| 5 | 3 | 0.3 | 2 | 1.5 | 2 |

$$\Phi(K_w) = \sum_{i=1}^{N_s} e_i^2 \quad (5.6)$$

where $K_w = 1 \times N_p$ vector of the wall demand coefficients
$e_i$ = the difference between the *i*th pair of measured ($c_m$) and modeled ($c_o$) chlorine concentration
$N_s$ = number of samples (measurements)
$N_p$ = number of wall demand coefficients to be estimated

Data collection in DWDS will not allow estimation of a unique $K_w$ for each pipe. Rather, the pipes should be grouped and those in the same group are assumed to have the same wall demand coefficient. $G_i$ ($1 \leq I \leq N_p$) represents the sets of pipe indices that share the same wall demand coefficient $K_{wi}$.

Chlorine first-order decay is a key assumption here, and the linear relationship between upstream and downstream chlorine concentration can not be established without this assumption. Other assumptions are indicated below:

- During the process of wall demand coefficient estimation, the hydraulic model is assumed to be accurate; i.e., the hydraulic model is calibrated before the estimation of wall demand coefficients.
- Other parameters in the chlorine decay model, such as bulk decay coefficient, mass transfer coefficient and hydraulic radius are assumed to be known or estimated independently before the calibration of wall demand coefficient. The proposed estimation method is to calibrate only one kind of water quality model parameter: wall demand coefficient.
- Pipes are assumed to be grouped accurately based on their wall demand coefficient; i.e., pipes in the same group have the same true values of the wall demand coefficients.

*Multi Input Single Output Water Quality Model for Wall Demand Coefficient*

In the problem of estimating wall demand coefficient, the I/O model is used to establish the analytical relationship between input and output concentrations. As mentioned in the last section, the inputs can be any node upstream of the output as long as the I/O model is closed. One kind of possible inputs common in DWDS is water storage tanks. In water quality modeling, tanks may be assumed to be completely mixed, partially mixed, or plug flow. In reality, however, there are many types of tank geometries and operating characteristics, and tank modeling remains an uncertain element of a DWDS water quality model. Because the objective of the research is to calibrate pipe-wall demand coefficients, it is ideal to avoid tank modeling error. Therefore, if water reaching the output comes through one or more tanks, all these tanks are treated as input locations and chlorine concentrations are sampled at the tank outlets. Because the tanks are excluded from the wall demand coefficients, decay coefficients in tanks do not need to be determined.

Because wall demand coefficients are to be estimated, the PBA that constructs the I/O model must be revised to collect detailed travel path information for each backtracking particle. Each travel path j is described by a sequence of travel segment $TS_{j,l}$ such that segment l is associated with flow in a single pipe under constant hydraulic conditions. Thus first-order decay coefficient K within a travel segment is constant. Let $K_{j,l}$ and $t_{j,l}$ be defined as the decay coefficient and travel time in travel segment $TS_{j,l}$, respectively, and define $p_{j,l}$ as the pipe index for travel segment $TS_{j,l}$. If $p_{j,l}$ $G_i$, then travel segment $TS_{j,l}$ is associated with the grouped wall demand coefficient $K_{wi}$. The network shown in Figure 5.10 has one input I, one output O, and two pipes 1 and 2. Flow direction in the two pipes is always from I to O. The flow velocity, and therefore overall chlorine decay coefficient K, changes in the pipes, caused by water demand change at O. Two flow paths exist between the input and output. For a single output time T1, water particles reaching O split according to flow rates in pipe 1 and 2, and both of the particles a and b are backtracked. At time T2 (T2 < T1 because the algorithm is backtracking), hydraulic conditions in the pipes change, and the first travel segment in both flow paths temporarily ends. Under new hydraulic conditions, a and b are backtracked and b first reaches input I at time T3, ending one flow path and its second travel segment. At time T4, hydraulic conditions change again, and another new travel segment is generated for the remaining flow path, which stops at time T5 and node I. The I/O model constructed by PBA then includes one input and two flow paths: one has 2 travel segments and the other has 3. The effect of wall demand coefficients on output chlorine concentration at time T1 is reflected in the overall decay coefficients for the 5 travel segments. Because there are no tanks in the flow paths, the impact coefficient of each path can be expressed explicitly as a function of travel segment chlorine decay coefficients and travel times as shown in Equation 5.7.

$$\gamma_j = f_j \prod_{l \in P_j} \exp(-K_{j,l} t_{j,l}) = f_i \exp\left( \sum_{l \in P_j} -K_{j,l} t_{j,l} \right) \tag{5.7}$$

where $P_j$ = set of travel segment indices that constitute travel path j between the output and the inputs (obtained using the PBA)

$f_j$ = dilution factor equaling to the flow ratio of path j

When $p_{j,l,} \in G_i$,

$$K_{j,l} = K_b + \frac{K_{f,j,l}K_{wi}}{R_{H,j,l}(K_{f,j,l} + K_{wi})} \tag{5.8}$$

where $R_{H,j,l}$ = hydraulic radius of the travel segment $TS_{j,l}$
$K_{f,j,l}$ = mass transfer coefficient of the travel segment $TS_{j,l}$ respectively

Note that these quantities are known from the hydraulic simulation and the PBA. Combining Equations 5.5 and 5.7, a multiple-input/single-output model of chlorine concentration can be expressed as:

$$c_o(T) = \sum_{j=1}^{N} \gamma_j c_{ij}(T - t_j) = \sum_{j=1}^{N} f_j \exp\left(\sum_{l \in P_j} -K_{j,l}t_{j,l}\right) c_{ij}(T - t_j) \tag{5.9}$$

Notably, equations 5.8 and 5.9 express the model output explicitly in terms of pipe-wall demand coefficients, and an iterative nonlinear parameter estimation method can be implemented to calibrate the coefficients efficiently. The difference between the PBA revised for wall demand coefficient estimation and that described in Shang et al. (2002) is that the former collects information necessary for the calculation of path impact coefficients, while the latter accumulates those coefficient values during the particle backtracking process.

***Gauss-Newton Method for Parameter Estimation***

The Gauss-Newton method (Bard 1974) is proposed to solve the least square problem in Equation 5.6, with equations 5.8 and 5.9. The *i*th iteration of the method is:

$$\mathrm{K}_w(i + 1) = \mathrm{K}_w(i) - \mathrm{N}(i)^{-1}\mathrm{q}(i) \tag{5.10}$$

where N = approximated Hessian matrix H of the function $K_{wi}$
q = gradient vector of the function $\Phi(K_w)$

Each element of the Hessian matrix is defined as:

$$H_{\alpha\beta} = \frac{\partial^2 \Phi}{\partial K_{w\alpha} \partial K_{w\beta}} = 2\sum_{i=1}^{N_s} e_i \frac{\partial^2 e_i}{\partial K_{w\alpha} \partial K_{w\beta}} + 2\sum_{i=1}^{N_s} \left(\frac{\partial e_i}{\partial K_{w\alpha}}\right)\left(\frac{\partial e_i}{\partial K_{w\beta}}\right)$$

In the Gauss-Newton method, the first term of $H_{\alpha\beta}$ is neglected and N is used to approximate H, where H is defined

$$N_{\alpha\beta} = 2\sum_{i=1}^{N_s}\left(\frac{\partial e_i}{\partial K_{w\alpha}}\right)\left(\frac{\partial e_i}{\partial K_{w\beta}}\right)$$

Each element of the gradient vector is defined

$$q_\alpha = \frac{\partial \Phi}{\partial K_{w\alpha}} = 2\sum_{i=1}^{N_s}\left(\frac{\partial e_i}{\partial K_{w\alpha}}\right)$$

***Model Sensitivity Analysis***

In the description of the iterative Gauss-Newton method, the derivative of the difference between measured and modeled chlorine concentration e with respect to wall demand coefficient needs to be calculated.

From the definition of e and Equation 5.9,

$$\frac{\partial e(t)}{\partial K_{wi}} = -\frac{\partial c_o(T)}{\partial K_{wi}} = -\sum_{j=1}^{N}\frac{\partial \gamma_j}{\partial K_{wi}}c_j(T-t_j) \tag{5.11}$$

From Equation 5.7

$$-\frac{\partial \gamma_j}{\partial K_{wi}} = \gamma_j \sum_{l \in P_j} t_{j,l}\frac{\partial K_{j,l}}{\partial K_{wi}} \tag{5.12}$$

Combine Equations 5.11 and 5.12

$$\frac{\partial e(t)}{\partial K_{wi}} = \sum_{j=1}^{N}\gamma_j \cdot \left(\sum_{l \in P_j} t_{j,l}\frac{\partial K_{j,l}}{\partial K_{wi}}\right) \cdot c_j(T-t_j) \tag{5.13}$$

From the definition of overall decay coefficient (Equation 5.4),

$$\frac{\partial K_{j,l}}{\partial K_{wi}} = \begin{cases} \dfrac{K_{f,j,l}^2}{R_{H,j,l}(K_{f,j,l}+K_{wi})^2} & \text{for } p_{j,l} \in G_i \\ 0 & \text{for } p_{j,l} \notin G_i \end{cases} \tag{5.14}$$

Whenever the estimates of wall demand coefficients are updated, the model sensitivity analysis is performed, because $\partial e(T)/\partial K_{wi}$ depends on $K_{wi}$. All the parameters in Equation 5.13 and 5.14 other than $K_{wi}$ are collected during the particle backtracking process. Therefore, the PBA is required only once for each measurement, and the update of model sensitivity is computationally efficient.

**Estimate Uncertainties Analysis Under Chlorine Measurement Error**

Both chlorine measurements and the hydraulic model that is the base of water quality modeling may contain errors. Hydraulic model calibration is not perfect, and resulted models contain errors in different hydraulic data, such as water demand, pipe roughness, and storage tank level. These kinds of errors will propagate to pipe flow velocity, and the simulation of chlorine transport within the DWDS may differ from the real world. The impact of hydraulic error on wall demand coefficient estimate was considered in the sampling design process, which is discussed in the model calibration section.

The uncertainties of parameter estimates caused by measurement errors on model output variables have been discussed widely in parameter estimation literature. When the output measurement errors have independent and identical distribution (IID), the estimate uncertainty matrix can be approximated according to Equation 5.15 (Bard 1974).

$$V_o = 2\sigma^2 N^{-1} \tag{5.15}$$

where $V_o$ = variance matrix of the estimates only under output measurement error
$\sigma^2$ = variance of chlorine measurement error

In the I/O model, not only output concentration but also input concentration contain measurement error, and the estimate uncertainties caused by measurement errors at inputs need to be quantified.

Under estimated $K_w$, objective function $\Phi(K_w)$ reaches a local minimum. When input concentration measurements vary slightly from $c_i$ to $c_i + \delta ci$, where $c_i$ is $1 \times N_{ci}$ vector that contains input concentrations having impact on output water quality, minimization of $\Phi(K_w)$ causes the estimates to change to $K_w + \delta K_w$.

$$\frac{\partial \Phi(c_i, K_w)}{\partial K_w} = 0 \tag{5.16}$$

$$\frac{\partial \Phi(c_i + \delta c_i K_w + \delta K_w)}{\partial K_w} = 0 \tag{5.17}$$

Equation 5.17 is expanded in Taylor series, and terms up to first-order subtracting Equation 5.4 leads to

$$\frac{\delta^2\Phi}{\delta K_w^2}\delta k_w + \left(\frac{\delta^2\Phi}{\delta K_w \delta c_i}\right)\delta c_i \approx 0 \tag{5.18}$$

Therefore,

$$\delta K_w = -H^{-1}\left(\frac{\delta^2\Phi}{\delta K_w \delta c_i}\right)\delta c_i \tag{5.19}$$

where the Hessian matrix is H is approximated as N. The estimate covariance matrix V is

$$V \equiv E(\delta K_w \delta K_w^T) \tag{5.20}$$

Combining equations 16 and 17, the estimate variance only under input measurement errors $V_i$ is

$$V_i \approx E\left(N^{-1}\left(\frac{\partial^2\Phi}{\partial K_w \partial c_i}\right)\delta c_i \delta c_i^T\left(\frac{\partial^2\Phi}{\partial K_w \partial c_i}\right)^T N^{-1}\right) \tag{5.21}$$

When measurement errors on input concentration also have IID distribution with variance $\sigma^2$,

$$V_i = \sigma^2 N^{-1}\left(\frac{\partial^2\Phi}{\partial K_w \partial c_i}\right)\left(\frac{\partial^2\Phi}{\partial K_w \partial c_i}\right)^T N^{-1} \tag{5.22}$$

Define $f_\beta$ to be the $\beta$th element in input concentration vector $c_i$ and matrix L to be

$$L = \frac{\partial^2\Phi}{\partial K_w \partial c_i} \tag{5.23}$$

Then

$$V_i = \sigma^2 N^{-1} L L^T N^{-1} \tag{5.24}$$

The element $L_{\alpha\beta}$ at row $\alpha$ and column $\beta$ of matrix L is

$$L_{\alpha\beta} = \frac{\partial^2\Phi}{\partial K_{w\alpha}\partial K_{w\beta}} = 2\sum_{i=1}^{N_S} e_i\left(\frac{\partial^2 e_i}{\partial K_{w\alpha}\partial f_\beta}\right) + 2\sum_{i=1}^{N_S}\left(\frac{\partial e_i}{\partial K_{w\alpha}}\right)\left(\frac{\partial e_i}{\partial f_\beta}\right) \approx 2\sum_{i=1}^{N_S}\left(\frac{\partial e_i}{\partial K_{w\alpha}}\right)\left(\frac{\partial e_i}{\partial f_\beta}\right) \tag{5.25}$$

The element $LL^T_{xy}$ at row x and column y of matrix $LL^T$ is then

$$(LL^T)_{xy} = 4\sum_{\beta=1}^{N_{c_i}}\left[\sum_{i=1}^{N_S}\left(\frac{\partial e_i}{\partial K_{\omega x}}\right)\left(\frac{\partial e_i}{\partial f_\beta}\right)\sum_{i'=1}^{N_S}\left(\frac{\partial e_{i'}}{\partial K_{\omega y}}\right)\left(\frac{\partial e_{i'}}{\partial f_\beta}\right)\right] \tag{5.26}$$

Elements of the vector $c_i$ are chlorine concentrations sampled at input locations at different times. When input concentrations are sampled continuously, the sensitivity of output concentration with respect to input concentration equals the impact coefficient of corresponding flow path. When the sampling frequency at outputs is much lower than that at inputs, it can be reasonable to assume that input concentration at given time and input location will affect only one sampled output concentration. This assumption excludes the possibilities that water leaving an input sometimes reaches the same output at different sampling times or different output locations at the same or different sampling times. This assumption simplifies Equation 5.26 because

$$\frac{\partial e_i}{f_\beta}\frac{\partial e_{i'}}{f_\beta} = 0$$

when $i \neq i'$.

$$(LL^T)_{xy} = 4\sum_{i=1}^{N_S}\left(\frac{\partial e_i}{\partial K_{\omega x}}\right)\left(\frac{\partial e_i}{\partial K_{\omega y}}\right)\sum_{\beta=1}^{N_{c_i}}\left(\frac{\partial e_i}{\partial f_\beta}\right)^2 \tag{5.27}$$

In Equation 5.27,

$$\sum_{\beta=1}^{N_{c_i}}\left(\frac{\partial e_i}{\partial f_\beta}\right)^2$$

equals to the sum of the squared impact coefficients for the output concentration $c_o$ corresponding to $e_i$.

Equations 5.24 and 5.27 together provide an easy way to calculate the estimated uncertainties caused by measurement errors on input concentration. Assuming input and output measurement errors are not correlated, the estimate variance matrix under both kinds of error is then,

$$V = V_o + V_i \tag{5.28}$$

The calculation of $V_o$ and $V_i$ requires impact coefficients of flow paths and sensitivities of modeled output concentration to wall demand coefficients. Both pieces of information can be obtained together with parameter estimates based on calibrated wall demand coefficients.

**Table 5.2**
**Simulation results for wall demand coefficient estimation**

| Design | $\mu_1$ (ft/day) | $\mu_2$ (ft/day) | $\sigma_1$ (ft/day) | $\sigma_2$ (ft/day) | $\widehat{\sigma_1}$ (ft/day) | $\widehat{\sigma_1}$ (ft/day) | $\mu_r$ (mg/L) | $\sigma_r$ (mg/L) |
|---|---|---|---|---|---|---|---|---|
| A | 1.50 | 2.51 | 0.024 | 0.053 | 0.023-0.025 | 0.051-0.055 | 0.053 | 0.003 |
| B | 1.50 | 2.51 | 0.030 | 0.055 | 0.030-0.031 | 0.055-0.059 | 0.056 | 0.003 |
| C | 1.50 | 2.50 | 0.036 | 0.073 | 0.034-0.036 | 0.069-0.074 | 0.051 | 0.003 |

## Application

The proposed parameter estimation approach to calibrate pipe-wall demand coefficients is tested in a real network shown in Figure 5.11. Node S, which is the treatment plant, provides water continuously to the network with flowrate from 2375 to 3790 gallons per minute. There are six elevated tanks (T1 to T6) and 4 pumps in the network, and the filling and draining of the tanks are controlled by the operation of pumps according to the water levels in the tanks. A hydraulic model of the network was calibrated for a period of 16 days (384 hours) using pressure measurements, tank water levels, and water meter readings. Water leaving the treatment plant, S, was measured for free chlorine concentration and the data are plotted in Figure 5.11. In the simulation study, pipes in the application network are assigned to belong to one of the two pipe groups. These pipes bolded in the figure (group 1) are assumed to have true wall demand coefficient 2.5 ft/day, while the other pipes are grouped together (group 2) with wall demand coefficient 1.5 ft/day.

Three designs (A, B, and C) are tested for wall demand coefficient estimation. In each design, chlorine concentrations are measured at two output locations (X1 and X2 for design X) and the corresponding input locations (S and the tanks that are in the flow path from S to the output locations). As mentioned earlier, sampling chlorine concentration at the tanks can exclude tank modeling error from wall demand coefficient estimation. Chlorine is sampled at the outputs from time 301 to time 324 with frequency 4 times an hour. Therefore, there are 96 chlorine measurements at each output location. The sampling durations at the input locations (S and tanks) are assumed to be long enough to close the I/O relationship between any output measurement and input ones. In other words, all the water reaching the output from time 301 to 324 must be measured for chlorine concentration when leaving inputs. It is assumed that the hydraulic model does not contain any error. Both of the input and output chlorine measurements are assumed to be subject to errors with IID. The standards deviation of the IID error added to chlorine measurements is 0.05 mg/L. Monte Carlo simulation with 500 realizations is implemented for each design to study the uncertainties in parameter estimates, and the results are shown in Table 5.2, where $\mu_i$ and $\sigma_i$ represent the mean and standard deviation of the wall demand coefficient estimate for pipe group i respectively. The $\widehat{\sigma_i}$ is the estimated standard deviation for parameter i and reflects the uncertainty in wall demand coefficient estimation caused by the assumed chlorine measurement errors. The calculation of $\widehat{\sigma_i}$ is described in last section, and $\widehat{\sigma_i} = V_{ii}$, where Vii is the diagonal elements of matrix V in Equation 5.28. Model performance to fit the sampled data is described by the mean $\mu_r$ and standard deviation $\sigma_r$ of the root mean square error (RMSE). Both $\mu_r$ and $\sigma_r$ are with unit mg/L.

**Table 5.3**
**Simulation results for bad design**

| Output locations | $\mu_1$ (ft/day) | $\mu_2$ (ft/day) | $\sigma_1$ (ft/day) | $\sigma_2$ (ft/day) | $\mu_r$ (mg/L) | $\sigma_r$ (mg/L) |
|---|---|---|---|---|---|---|
| A1, B1 | 1.55 | 2.50 | 0.57 | 0.29 | 0.053 | 0.003 |

***Discussion of the Application***

For the application example, three designs produce different estimate uncertainties, and the results show that sampling locations need to be selected carefully in the design period. In fact, all of the three designs are relatively "good" designs selected using some metric that is based on sensitivities analysis. The sampling design metric is the determinant of the information matrix, which is the approximated Hessian matrix N discussed in the parameter estimation section. Based on the D-Optimum principle (Fedorov 1972), designs with a large determinant of information matrix are "good" ones, while designs with a small determinant of information matrix will result in uncertain estimates. The following example can illustrate that a "bad" design for the application network can generate wall demand coefficients estimates with such high uncertainties that the estimates can hardly be used in chlorine residual modeling.

From Table 5.2, it can be seen that design A is the best of the three. This design is also the optimal design selected by D-optimum criteria if output locations are set to be two and the sampling duration and frequency are restricted to be the same as those of design A. Now change one output location from A2 to B1 and the Monte-Carlo simulation results are shown in Table 5.3. For this design, the standard deviation (0.57 ft/day) of the estimates for wall demand coefficient of pipe group 1 more than one third of the true value (1.50 ft/day).

It can be observed that the uncertainties contained in wall demand coefficient estimate can not be determined based on the fitness of mode output to observed data. In both good and bad designs in the application, the RMSE has mean about 0.05 mg/L and very low variability. However, the uncertainties in the parameter estimates can be estimated very well through the sensitivity analysis discussed previously. All the information that is needed to calculate the are collected in the parameter estimation algorithm. Because the water quality model is nonlinear in terms of wall demand coefficients, the estimate variance matrix V and values of depend on the estimated wall demand coefficients. It is obvious that the parameter uncertainties can not be estimated accurately when the estimates are far from the true values.

**Conclusions**

A novel approach to calibrate pipe-wall demand coefficients of water distribution systems is proposed and tested in simulation analysis. Under the assumption of first-order chlorine decay, the PBA algorithm is used to provide the detailed travel path information between input and output locations. With such information, the modeled chlorine output can be expressed explicitly in terms of the wall demand coefficients, and, therefore, the sensitivity of the modeled chlorine concentration to demand coefficients can be calculated efficiently. Nonlinear least-square estimation with Gauss-Newton iteration is used to identify the optimal estimates. The iterative estimation is efficient because the model outputs' sensitivity to parameters to be estimated is explicit. Therefore input-output water quality modeling by PBA is required only once for each modeled output in the estimation process. Another advantage of having the sensitivities is that the uncertainties of the wall demand coefficient estimates can be reliably estimated if the estimates are close to the true values. An application for a real network shows that large uncertainties can be caused by chlorine measurement errors and output locations are very important to get reliable estimates. Sampling design method is used in the application to select output locations under the assumption that the hydraulic model is completely accurate. The details of the sampling design that considers both chlorine measurement and hydraulic model error will be discussed elsewhere.

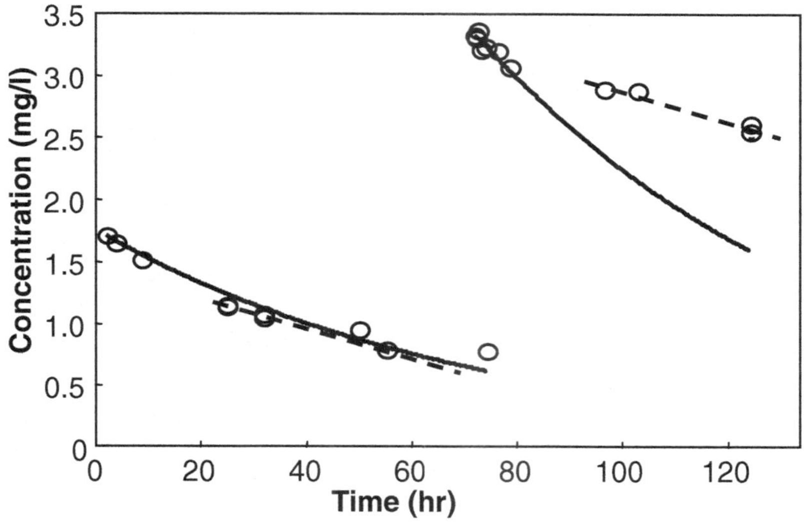


**Figure 5.1 Free chlorine decay and single-species (1st order) model error**

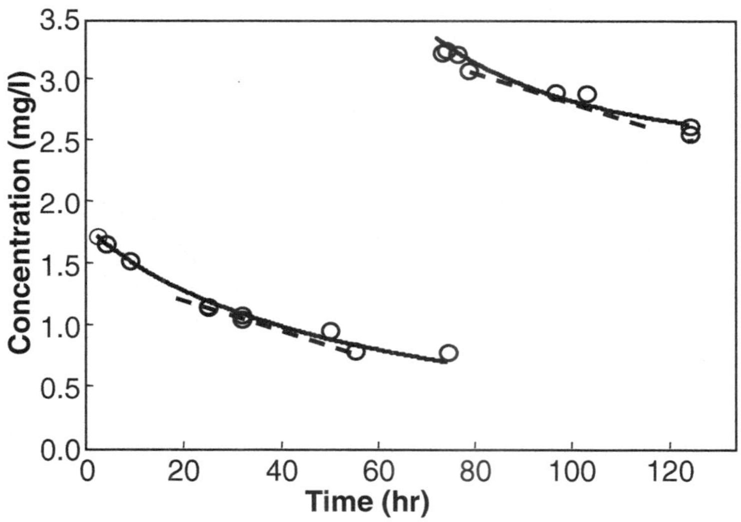


**Figure 5.2 Free chlorine decay and two-species model behavior**

**Figure 5.3 Mixing of water with different residence times in simple distribution system**

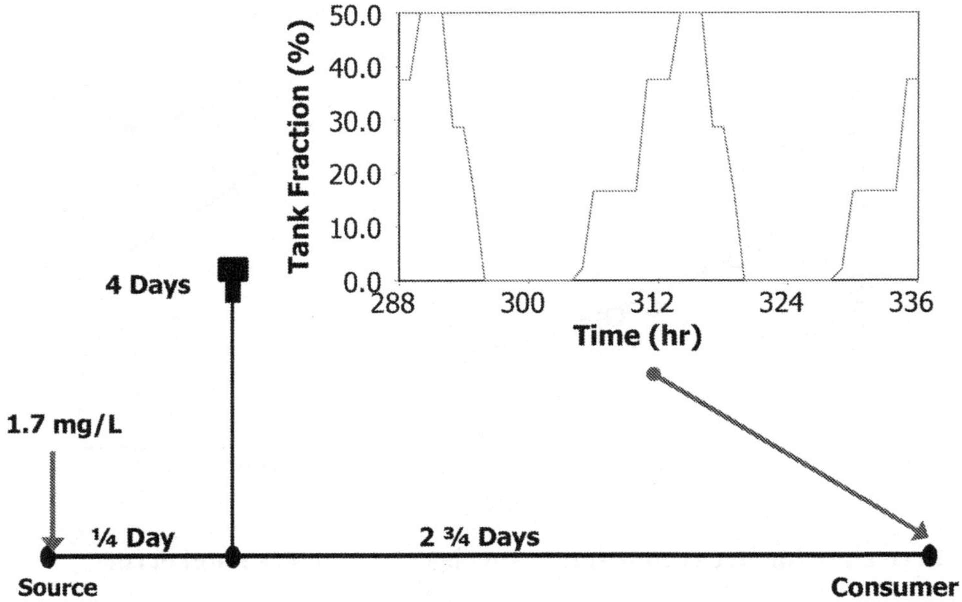


**Figure 5.4 Simple network illustrating potential differences between single- and two-species decay models (no rechlorination)**

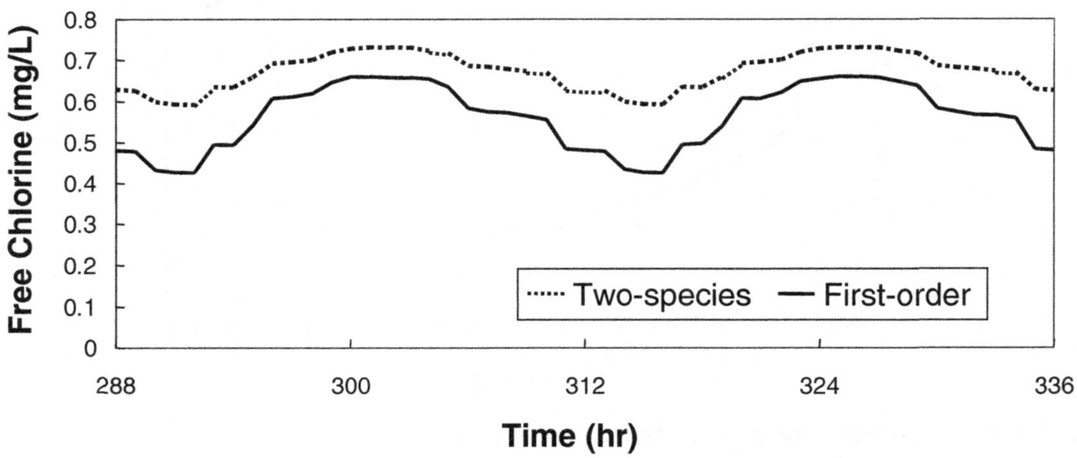


**Figure 5.5 Multi-species network simulation—comparison of single- and two-species chlorine decay models (no rechlorination)**

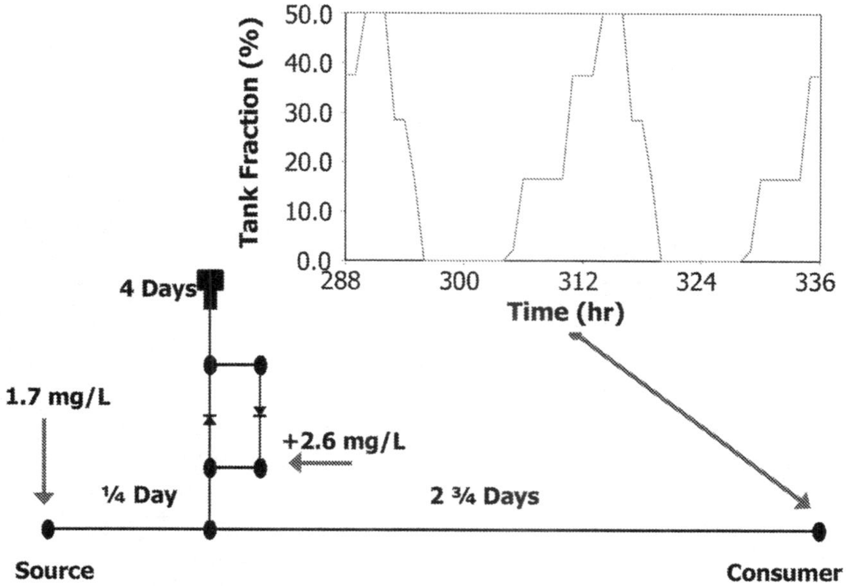


**Figure 5.6 Simple network example illustrating potential differences between single- and two-species decay models (with rechlorination)**

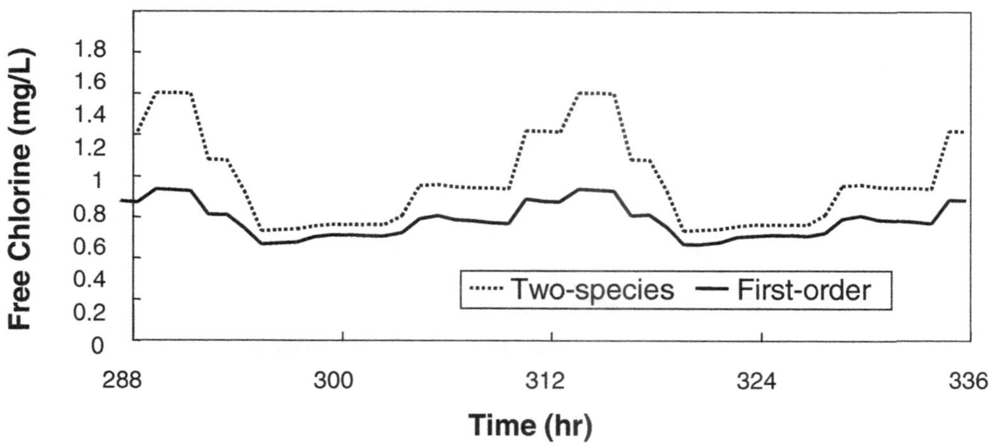


**Figure 5.7 Multi-species network simulation—comparison of single- and two-species chlorine decay models (with rechlorination)**

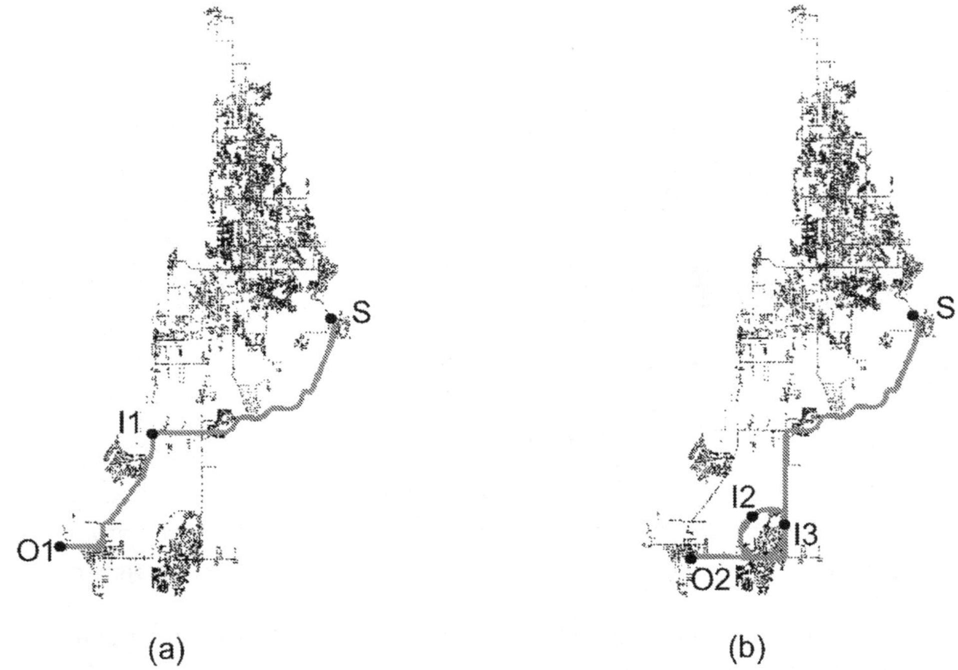


**Figure 5.8 Flow path for two sampling locations in a DWDS (only nodes are shown to increase visibility of pipe path)**

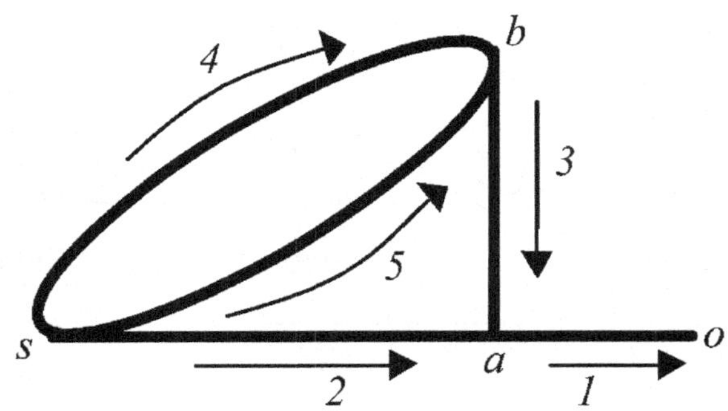


**Figure 5.9 Example for linear relationship between input and output**

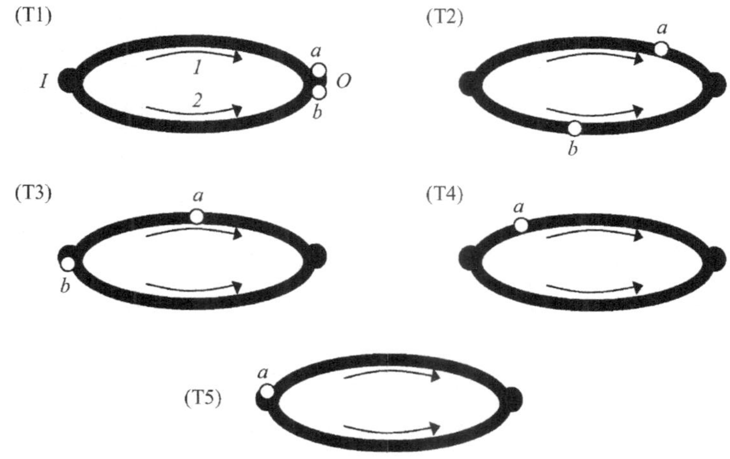


**Figure 5.10 Travel segments in flow path**

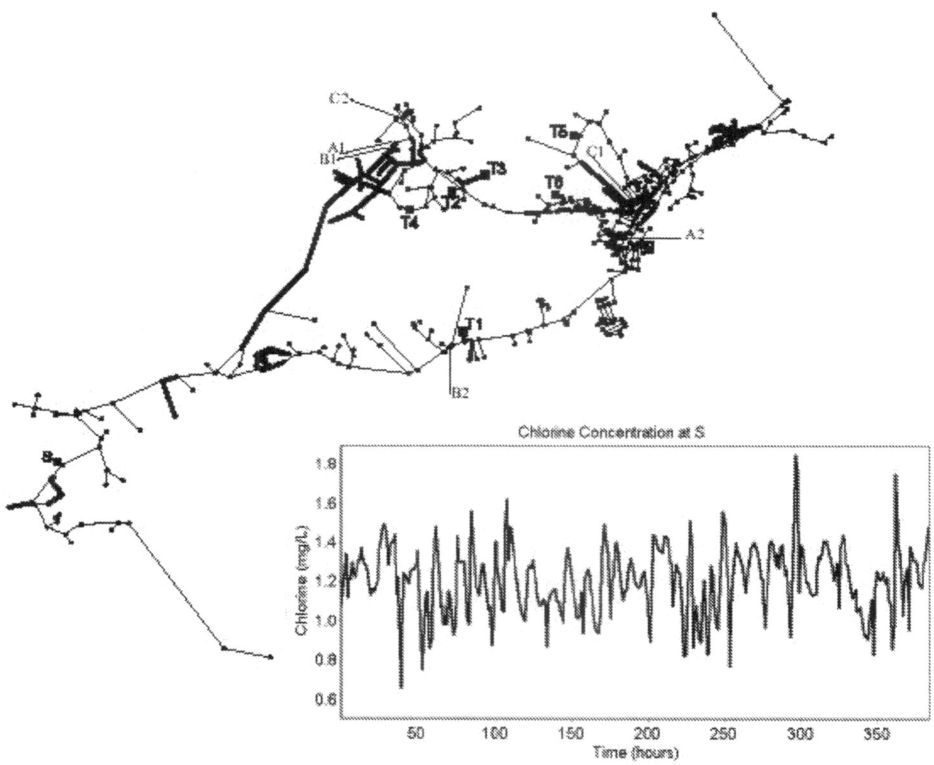


**Figure 5.11 Application network**

# CHAPTER 6
# FIELD-SCALE TESTING

Field data were collected from three utilities: the City of Newport News (Virginia), the City of Scottsdale (Arizona), and "Utility A," located in the southeastern United States. The most complete data set and detailed analysis are from Utility A. Methods for data collection from Newport News and Scottsdale are presented as alternative methods of collecting data that are less intensive, yet still provide valuable information with respect to chlorine and chloramine decay in the distribution system.

## CITY OF NEWPORT NEWS, VIRGINIA

The City of Newport News operates two WTPs, the Lee Hall WTP and the Harwood's Mill WTP. Both plants receive at least 75 percent of their raw water from the Chickahominy River. The treatment processes used at both plants are the same: coagulation, clarification, and granular media filtration. Primary disinfection is achieved at Lee Hall WTP with free chlorine, and ozone is used at Harwood's Mill. Both plants converted to chloramine as a secondary disinfectant in 1998. The ozone system was installed at Harwood's Mill in early 2002. Originally, the team planned to monitor changes in chloramine residual as a result of the change in treatment. However, the logistics of additional monitoring by the utility at a time when the system conversion was stressing their resources did not allow them to conduct the additional monitoring for this research. As such, monitoring was conducted after the conversion.

### Chloramine Residual Monitoring

The City of Newport News purchased and installed two chlorine/chloramine residual monitors. The location of the monitors is shown in Figure 6.1. The monitors were installed in late May 2002, but due to difficulties integrating with their data acquisition system, data were not available until August 2002. Figures 6.2 and 6.3 show total chlorine residual leaving the plant and at the analyzer in the distribution system for a period of four months (August 2002 – December 2002). Figure 6.2 shows data for Harwood's Mill WTP and Figure 6.3 shows data for Lee Hall WTP. The difference in observed decay on the two figures can be attributed to the difference in distribution system travel time.

Figure 6.4 shows box plots of the chlorine residual leaving each plant, the residual in the distribution system, and the observed chlorine decay. The average decay in the Harwood's Mill and Lee Hall portions of the system are 1.1 mg/L and 0.5 mg/L, respectively. The estimated travel time to each site is 46 hours for the Harwood's Mill monitoring site and 7.4 hours for the Lee Hall monitoring site. The water age information is based on hydraulic modeling performed by the City. The travel time and chloramine decay are summarized in Table 6.1.

## CITY OF SCOTTSDALE, ARIZONA

The City of Scottsdale currently produces water via surface water treatment, ground water production, and aquifer storage and recovery facilities. The surface water source is the Central Arizona Project (CAP) Canal, which transports water from the Colorado River across Arizona

**Table 6.1**
**Chloramine decay in Newport News distribution system**

| | Travel time | Chloramine decay (mg/L) |
|---|---|---|
| Harwood's Mill WTP to Poquoson | 46 hrs | 1.1 |
| Lee Hall WTP to Anheuser-Busch | 7.4 hrs | 0.5 |

from Lake Havasu City through the Phoenix metropolitan area to Tucson. The surface water treatment processes include alum coagualation, sedimentation, and filtration. Free chlorine is used as the primary and secondary disinfectant.

### Methods and Materials

A portion of the distribution system was selected for the chlorine decay study as shown in Figure 6.5. This area is in the eastern-most portion of Scottsdale in an area the City staff identifies as historically having problems maintaining a chlorine residual. The study area is located approximately 10 miles from the City's WTP. Most of the water entering the zone is transmitted through a 30-inch pipe that leaves the WTP and travels Southeast to Reservoir 114. Alternately, water is transmitted South from the WTP and then East along Shea Boulevard where is it pumped into Reservoir 114.

In order to provide hydraulic data necessary for model calibration in this area, a tracer study was performed during the week of September 23, 2002. Fluoride was added at Reservoir 114 for a period of 48 hours. The project team monitored the increase and decrease of fluoride in the study area. Three autosampling units were placed in the distribution system: two at private residences and one at Reservoir 36-5. The autosamplers collected samples every two hours for the duration of the test. Samples were retrieved from the units once per day and delivered to the City of Scottsdale Water Quality lab for analysis. The locations of the feed points and sampling points are shown in Figure 6.6.

### Results

The fluoride response curves at each site are shown in Figure 6.7. When calculating travel time, it was assumed that if fluoride had been fed longer, the system would have reached equilibrium and a maximum concentration of 1.27 mg/L would have been observed at each site. The travel time to each site was calculated by determining how long it took for fluoride to reach the mid-point between baseline conditions (0.33 mg/L) and the maximum concentration (1.27 mg/L). Data show the following travel times:

- 15 hours to Site A
- 24 hours to Site B
- 37 hours to Site C

The bumpiness seen in the fluoride data for Sites A and B shows the influence of storage tanks in the area. For example, Site A is very close to Reservoir 36-5 (at Site C). When Reservoir

**Table 6.2**
**Chlorine decay in Scottsdale distribution system**

| | Travel time | Chlorine decay (mg/L) |
|---|---|---|
| Site A | 15 hrs | 0.12 |
| Site B | 24 hrs | 0.32 |
| Site C | 37 hrs | 0.27 |

36-5 is filling, Site A receives water directly from the pumps at the fluoride feed location (Reservoir 114). However, when Reservoir 36-5 is draining, Site A receives water directly from Reservoir 36-5. Figure 6.8 shows the fluoride response data for Site A and Site C. The times of day when the reservoir is draining and Site A is receiving water from the Reservoir are circled. Excluding those points, the trend line shows the fluoride response at Site A if the only water source was the booster pumps at Reservoir 114. The dotted line shows what the data that are missing due to the equipment malfunction may look like.

Chlorine residual at several locations within the study area was monitored and recorded during the study using permanently installed chlorine analyzers and the City's SCADA system. An additional chlorine analyzer was obtained by the team, but it did not arrive in time to be installed at the sites during the tracer study. As such, it was planned that once the analyzer arrived, it would be installed at Site A and then moved to Site B.

Once the analyzer arrived in mid-October, the team worked with City staff to interface the analyzer with a datalogger and install the unit at the tracer test sites. Unfortunately, due to a faulty electrical switch in the analyzer, the 4 – 20 mA output was not received correctly by the datalogger. The chlorine analyzer was replaced by the manufacturer in early December. Chlorine data for Site A on 5-minute intervals are available for one week in December. After recording chlorine residual data at Site A, the analyzer was moved to Site B where it recorded chlorine residual data for 3 weeks. The chlorine residual at the feed location (Reservoir 114) and each of the sampling sites is shown in Figure 6.9. The resulting chlorine decay is summarized in Table 6.2 along with the travel time to each site.

## SOUTHEASTERN UNITED STATES UTILITY

A utility in the southeastern United States supported application of the Nwtrowk Path Analysis (NPA) methodology to a portion of their service area; a schematic of the study area piping is shown in Figure 6.10 (this area represents roughly one-half the treatment plant service area). Figure 6.10 also shows sampling locations (red circles), and the pipe size distribution with red (and thicker) lines representing the largest, and dark blue (and thinner) lines representing the smallest piping. Distribution system flow paths begin at the single treatment plant source (AA) and continue through two main transmission lines (AF and AG), leading to two residential/commercial districts (areas 1 and 2). Flow paths continue through area 2 to feed area 1, the ultimate output destination for the water quality study. Flow path splitting and recombination through areas 1 and 2 contribute to a progressively more complex set of contributing flow paths as compared to locations AF and AG.

The treatment plant uses groundwater from a well field that is screened at 600- to 1200-foot depths. The raw water contains approximately 2.7 mg/L $H_2S$ (as sulfide). Approximately 40 percent of raw water is pumped to aeration towers, which strip approximately 5 percent of the sulfide, and combined aerated and non-aerated streams undergo further biological sulfide removal in two sequential 5-million-gallon baffled tanks. The sulfur oxidizing bacteria settle in the tanks or enter a 5-million-gallon baffled clearwell where the effluent free chlorine concentration was maintained at approximately 3.5 mg/L during the study period. The treatment train does not include filtration; turbidity levels range from 1 to 6 NTU in the clearwell effluent, and from 0.2 to 1.5 NTU at the distribution system point of entry.

The study area drops to sea level at its western boundary, compared to approximately 75 feet above sea level at the treatment plant. Although the northern service area (not shown) rises to a maximum of 130 feet above sea level, the system includes no floating storage, relying mainly on variable speed high service pumps (controlled by discharge pressure) to satisfy instantaneous demands and maintain system pressure. The system does include two pumped ground storage tanks, but they are operated infrequently and used as a supplement for fire demand. These two pumped storage tanks were off-line for the entire duration of the study, which was a request made by the team to simplify data interpretation and analysis.

Subareas 1 and 2 are shown in Figures 6.11 and 6.12, including the sampling locations within each area and the pipe size distribution. Sampling locations AH and AI monitor the two main flow paths through subarea 2, which in turn feed subarea 1 (determined by prior input-output analysis). Sampling locations AB and AE monitor the two flow paths feeding subarea 1, and AC and AD are interior locations that are affected by increasingly more complex sets of flow paths due to flow splitting and recombination as well as temporal variation caused by changes in water demand.

All sampling locations were located in the field at fire hydrants using standard hydrant adaptors, and a continuous flow rate of approximately 1.5 gpm was maintained to reduce residence time in the hydrant barrel. Each sampling location was monitored by a continuous conductivity and temperature analyzer housed in a security box and containing a grab sampling port (Figure 6.13). Sample locations AC, AD, and AE also were monitored by a continuous chlorine and pH analyzer, used to provide greater temporal resolution at these destination output locations (Figure 6.14). Grab samples were collected at all sampling locations and analyzed in the field for free chlorine and pH. Additional samples were collected for laboratory fluoride ion analysis (conducted on site). In addition to water quality sampling, hydraulic data were collected using a network of nine pressure loggers co-located with conductivity meters (using separate hydrants). An additional five pressure loggers distributed throughout the subarea 1 provided additional hydraulic grade line resolution that may help determine flow directions.

## Field Test Protocol

A saturated NaCl solution was added to the treatment plant finished water beginning at 11:30AM on 9 September 2003. The addition was designed to produce a series of 100-mg/L brine pulses after mixing, spaced in time over a 24-hour period. The 100-mg/L pulse was selected to more than double the background conductivity (400 μS/cm), and is a factor of three less than the maximum allowable NaCl increase based on applicable federal and state standards. Between pulses, the NaCl feed was turned off. During the addition and for an estimated 4-5 days residence time after the injection, the pulse arrival was to be recorded by the data-logging conductivity meters situated at hydrants and other access points. Other water quality measurements were made simultaneous with pulse arrival. A more detailed chronological protocol follows.

### *Brine Solution and Injection*

The USEPA secondary standard on chloride is 250 mg/L. The applicable primacy agency lists a primary standard on sodium of 160 mg/L. These standards reflect the ratio of molar weights of sodium to chloride (160/250 = 22.99/35.45 = 0.64). Based on historical data, the point of entry chloride concentration averages 18 mg/L with a maximum of 23 mg/L. Fewer data on sodium concentration are available, with a 2002 triennial report listing 10 mg/L sodium. Assuming conservative background sodium and chloride concentrations of 25 mg/L, the applicable standards limited a chloride concentration increase to 225/35.45 = 6:35 mMol/L, and the sodium concentration increase to 135/22.99 = 5.87 mMol/L. Evidently, the allowable sodium concentration controls the allowable NaCl addition to 5.87 mMol/L, or 343 mg/L.

Approximately 2200 gallons of saturated NaCl solution (350 g/L) was made using food-grade salt (Morton Culinox 999) and plant finished water. The solution was stored in a 2500-gallon reservoir (Figure 6.15). This volume was designed to produce at least three and perhaps four salt pulses over the 24-hour period, depending on actual plant production. Ninety-two 80-pound bags of salt were added to the reservoir, followed by filling with plant finished water. The saturated brine solution was mixed using a submersible pump discharging into 2.5-inch-diameter plastic piping that extended above the final water line and was redirected downward toward the surface of the brine solution. A significant amount of undissolved salt remained in the bottom of the reservoir after mixing overnight. The mixing pump was positioned near the brine solution outlet and allowed to operate continuously so that all solid NaCl was cleared from the outlet and would not enter the distribution system.

The brine solution was injected into the force main leaving the treatment plant that provides 100 percent of the study area water demand. Figure 6.16 shows the point of injection, which was just before a venturi meter. Brine solution was delivered through parallel positive displacement pumps that were flow paced using the flow calculated through the venturi meter (pumps are shown in Figure 6.15). The flow pacing was set up by plant staff to produce a constant mixed salt concentration of 100 mg/L up through a 14-million gallons per day (mgd) maximum total flow rate, based on a 350 g/L saturated brine solution concentration. A 4-gpm brine flow rate into a 20-mgd finished water discharge produced the target 100 mg/L salt pulse.

Figure 6.17 presents the designed conductivity concentrations during the 24-hour tracer injection period (time zero in Figure 6.17 is the start of injection). The design called for three pulses, but sufficient brine solution was left after the third pulse to create a fourth of approximately 2.5 hours duration. Simultaneous with the first salt pulse, the plant fluoride feed was shut down, and left off for 24 hours (through the final salt pulse). This change in operation was designed to produce a fluoride ion concentration drop from approximately 0.9 mg/L to the background level of 0.2 mg/L. It was discovered on site that in addition to the plant fluoride feed, fluoride was added at the groundwater wells. The actual fluoride concentration drop was considerably less as a result, from approximately 0.8 to 0.5 mg/L. Given the accuracy of the fluoride ion analysis method, this drop was sufficient to identify whether or not water had left the plant during the 24-hour injection period, but was less valuable as a quantitative measure of the degree of mixing in the distribution system (i.e., mixing of water leaving the plant during the injection period with water leaving the plant either before or after the injection period).

The project team was concerned about attenuation and loss of integrity of the conductivity pulses due to flow splitting and recombination in the looped areas of the distribution system. These processes are inevitable, but logically there is an optimum configuration of pulses that would provide maximum flow information. As an extreme example, a single pulse of infinite duration would provide flow and velocity information only during the single step change from background conductivity. Thus information is related to the excitation in the conductivity signal, yet too many or too frequent pulses might lose their distinctive signature, and require complex (and new) data analysis approaches. The field protocol was designed to use a range of pulse durations and spacing, so that the data might provide useful information for future tracer experiments.

### *Field Water Quality Measurements*

The study was designed to measure changes in water quality of the volume of water leaving the treatment plant during the 24 hour injection period. The conductivity and fluoride ion concentrations would provide near real-time evidence of water movement, and field water quality sampling would adapt by moving downstream, following the traced water volume. Prior simulation of distribution system residence times provided an estimate of the test duration and how the sampling network would be extended.

Coincident with start of the first salt pulse, water quality sampling for pH, temperature, free chlorine, and fluoride ion was initiated at the treatment plant and the two nearest distribution system water quality stations (AF and AG). Fluoride samples collected in the field were analyzed in the lab as a near real-time indicator of pulse arrival. In addition, conductivity data were collected from the continuous loggers periodically (once or twice per day), and plotted to visually show tracer movement.

Water quality sampling routes were extended downstream as the tracer pulse was observed at a station and were curtailed after all tracer was judged to have passed a station. In particular, the protocol called for extending to stations AH and AI once tracer was observed at AF, to AB once tracer was observed at AG, and to AE once tracer was observed at AH or AI. Once tracer had been observed at AB and AE, and passed AG, AH, and AI, the sampling network would concentrate on the four subarea 1 locations AB, AC, AD, and AE. The test was complete when all tracer had passed the most downstream stations AC and AD.

### *Bottle Testing*

In addition to field water quality measurement, two bottle tests were conducted using finished water leaving the plant during the 24 hour injection period. Bottles were filled without head space at the start of the first pulse, and approximately 12 hours afterward (mid-way through the injection period). The two tests were conducted to measure any changes in water quality reaction kinetics due to plant operation changes throughout the day. All bottles were stored at distribution system temperature in the dark, using a water bath that was continuously flushed with finished water. Bottles were periodically harvested and sampled to measure pH, temperature, and free chlorine concentration. A comparison of bottle and distribution system water quality changes is possible, and the bottle water quality data were used to determine bulk water reaction rate coefficients.

## Results

Figures 6.18 and 6.19 summarize the free chlorine, conductivity, and fluoride data. No adjustments to or filtering of the data have been made.

The continuous conductivity data show that the injection did more than double the background level, though the peak increase was somewhat less than the predicted 1000 μS/cm. The four tracer pulses were not significantly attenuated at stations AB, AF, and AG, as these lie along single flow paths from the treatment plant. Note, however, that the pulse duration and spacing can be lengthened or shortened by changes in distribution travel time caused by changes in system demand. The tracer arrived at stations AH and AI prior to arriving at the "upstream" station AF. Station AF is in the vicinity of two parallel mains, and apparently the flow paths from the treatment plant differ significantly between the two. Investigation of the piping and flow paths is ongoing, but the unexpected behavior at AF led to a sampling plan that missed the first part of the tracer movement through stations AH and AI.

The tracer was only moderately attenuated at stations AH and AI, which lie along principle flow paths through subarea 2 but are affected somewhat by looping within this subarea. Station AE, however, shows the obvious affects of flow splitting and recombination through subarea 2. Some of the peaks are significantly attenuated, due to blending of paths with different travel times (and thus different conductivity). The number of pulses and the pulse durations also reflect the characteristics of the main flow paths measured at AH and AI, plus unmeasured minor flow paths.

The tracer at stations AC, and especially AD, reflects more complex flow paths that nearly destroy any resemblance to the original set of pulses. While calibration of a hydraulic model to match the tracer signature at these locations will be a challenge, the signatures – and in particular their degradation – carry valuable information about the dynamics of flow paths, flow rates, and travel times in the distribution system.

## Network Path Data Analysis

Application of the NPA methodology requires a calibrated hydraulic model. Utility engineering staff have developed and regularly use a detailed hydraulic model of the study area supported by GIS infrastructure and water billing databases. Utility personnel initiated the development of an extended period hydraulic model in the 1980s and have maintained it continuously since that time. The model is used on a weekly basis to assess the impacts of proposed economic development. This high level of internal commitment to model development and use is consistent with utility confidence in the simulation results.

### *Hydraulic and Water Quality Model Calibration*

The first step in water quality model development and NPA application is to construct a suitably detailed, dynamic, hydraulic model. This step requires more data and careful calibration of model coefficients, compared to typical hydraulic analysis. The conductivity data collected from the field will be used to assess existing hydraulic model accuracy by comparing field conductivity measurement to model predictions. A second version of the model will be produced using base demand data from the September 2003 billing cycle and adjusting plant output over time to match records from the study period. Finally, a third version of the hydraulic model will be developed by varying the spatial demand allocation, or other factors as determined by model deficiencies.

Once hydraulic model errors have been quantified and reduced as much as practical, free chlorine decay kinetic coefficients will be obtained from the bottle test data, and the NPA methodology will be applied to estimate wall demand coefficients.

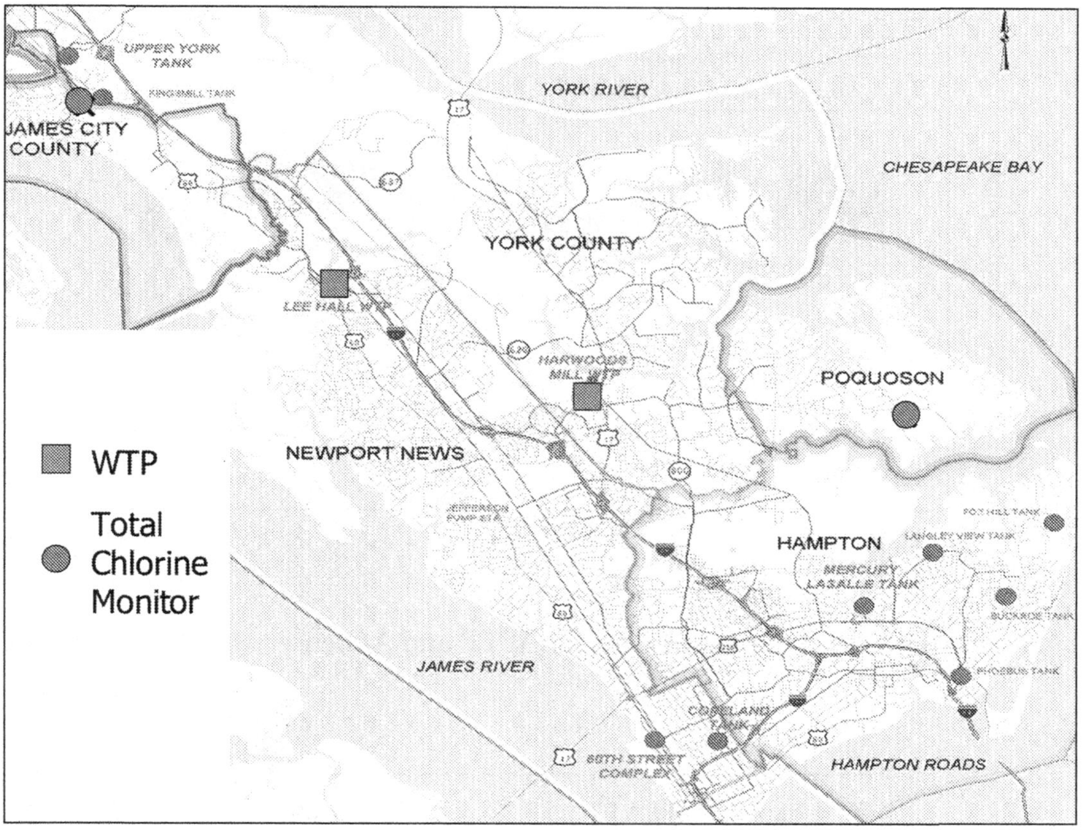


**Figure 6.1 Treatment plant and chlorine residual monitor locations in Newport News**

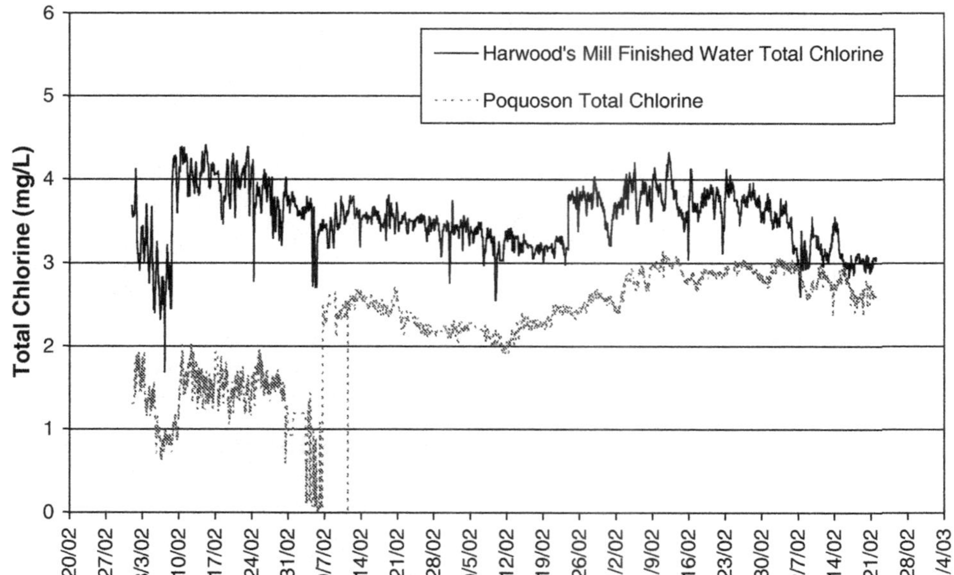


**Figure 6.2 Total chlorine residual at Harwood's Mill WTP and Poquoson chlorine analyzer**

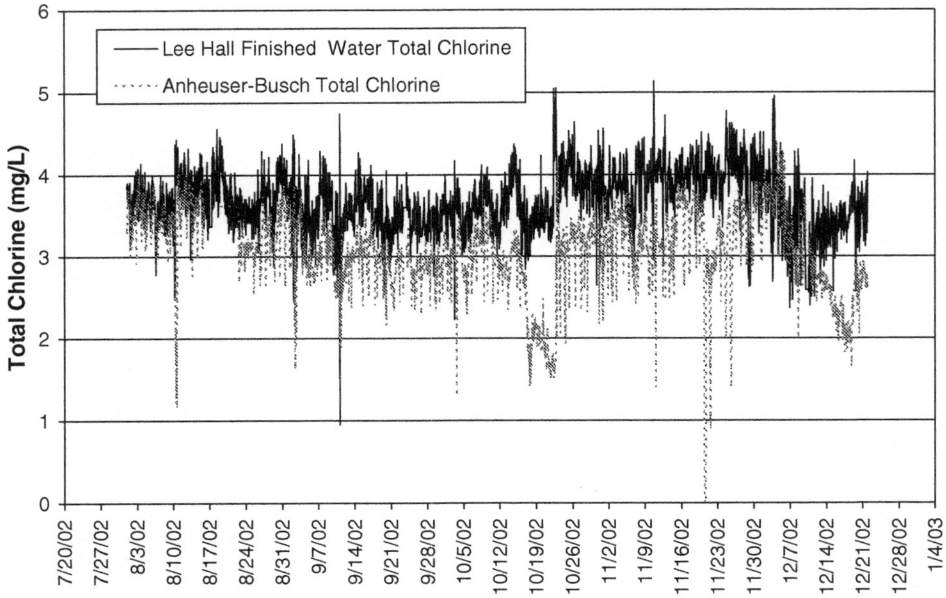


**Figure 6.3 Total chlorine residual at Lee Hall WTP and Anheuser-Busch chlorine analyzer**

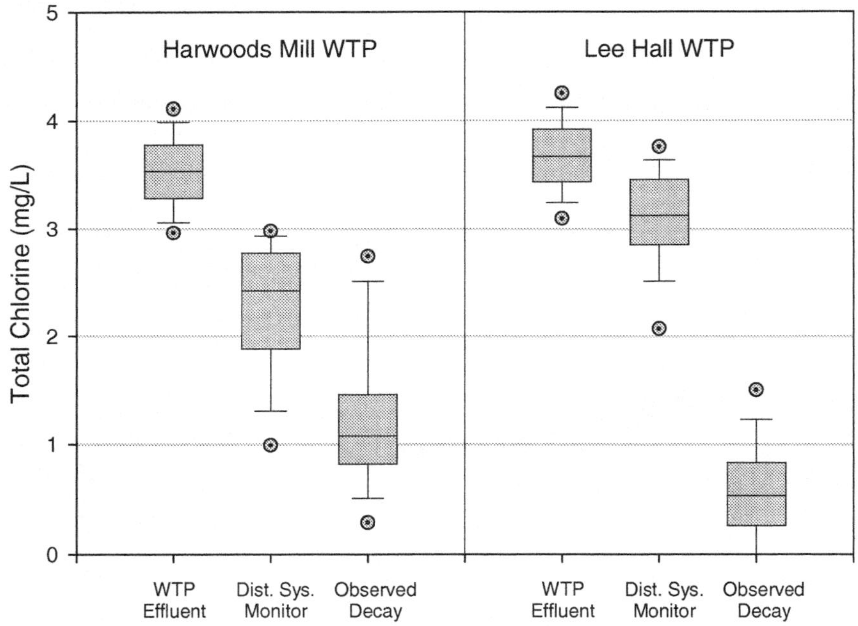


**Figure 6.4 Chlorine demand observed in each area of the distribution system**

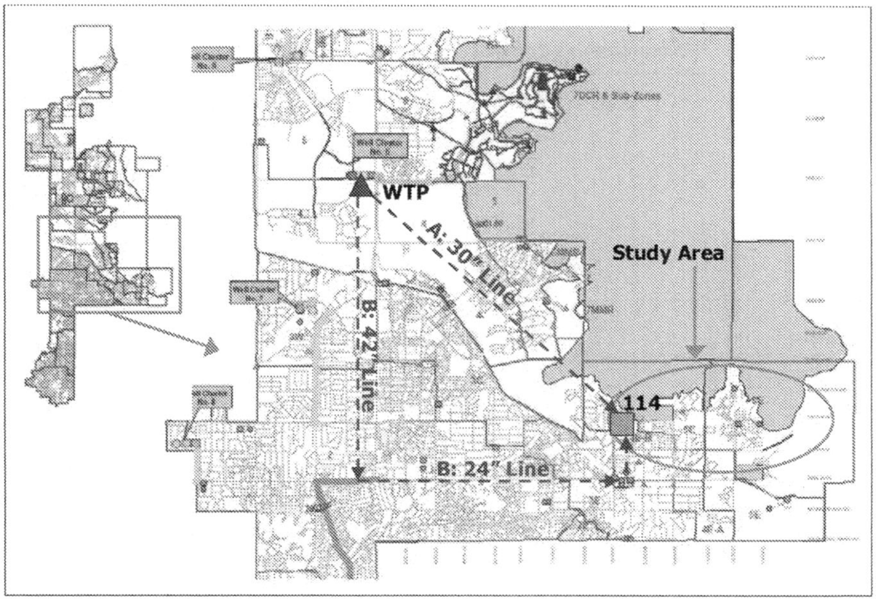


**Figure 6.5 Area selected for field testing in Scottsdale**

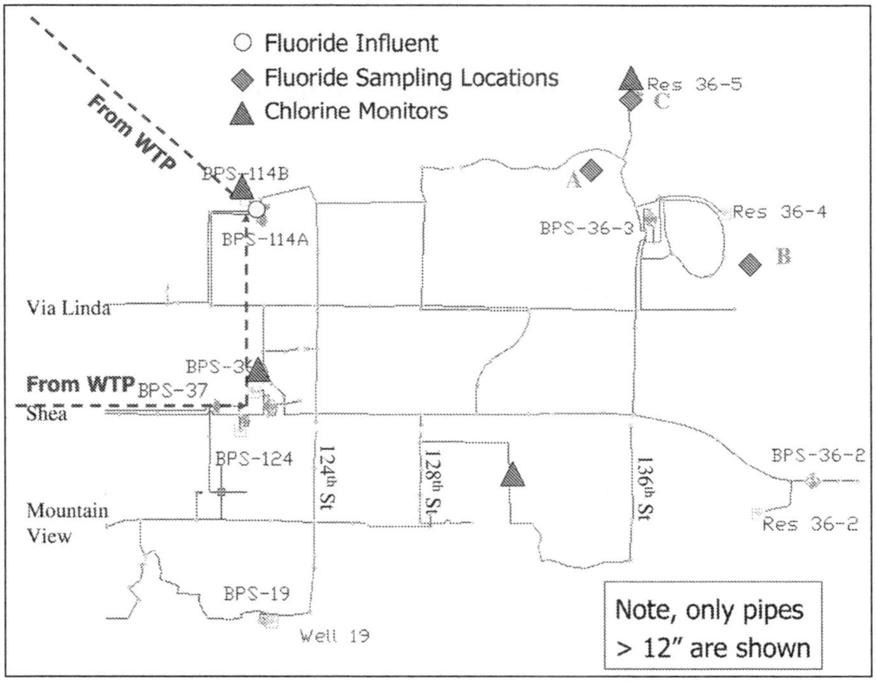


**Figure 6.6 Sampling locations for field testing in Scottsdale**

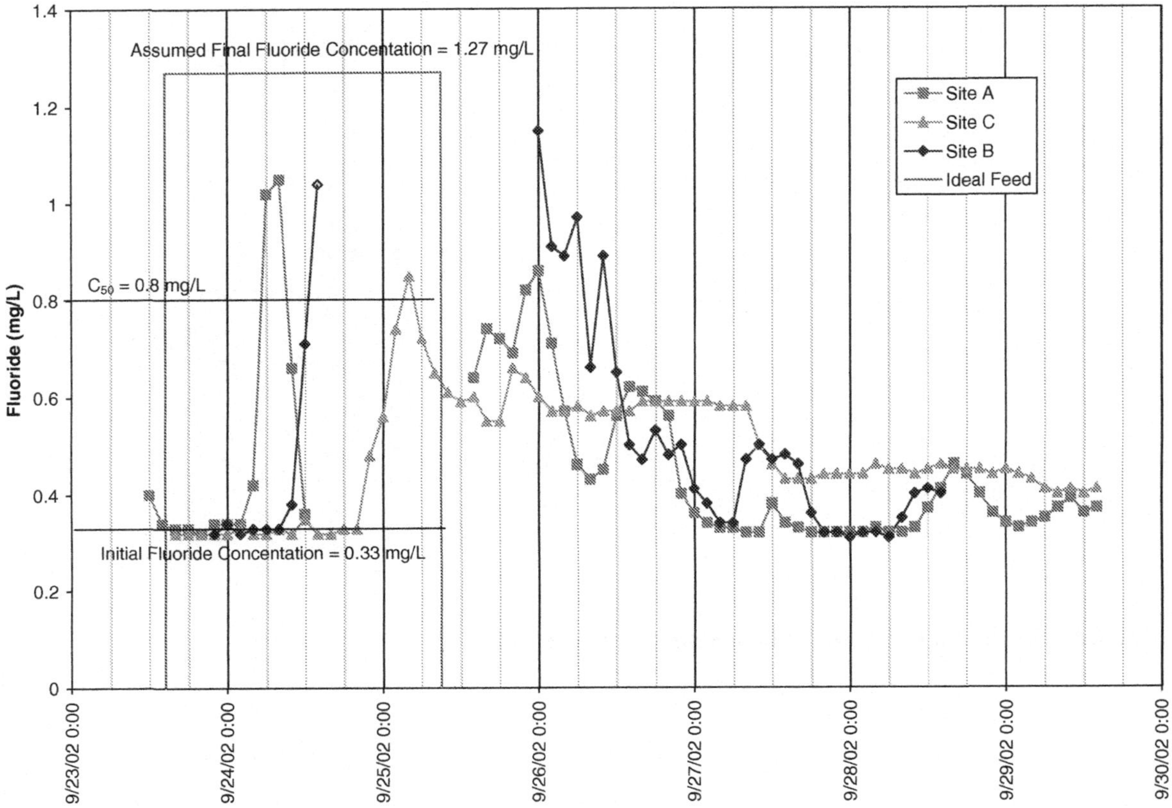


**Figure 6.7 Fluoride feed and response data for each sampling point**

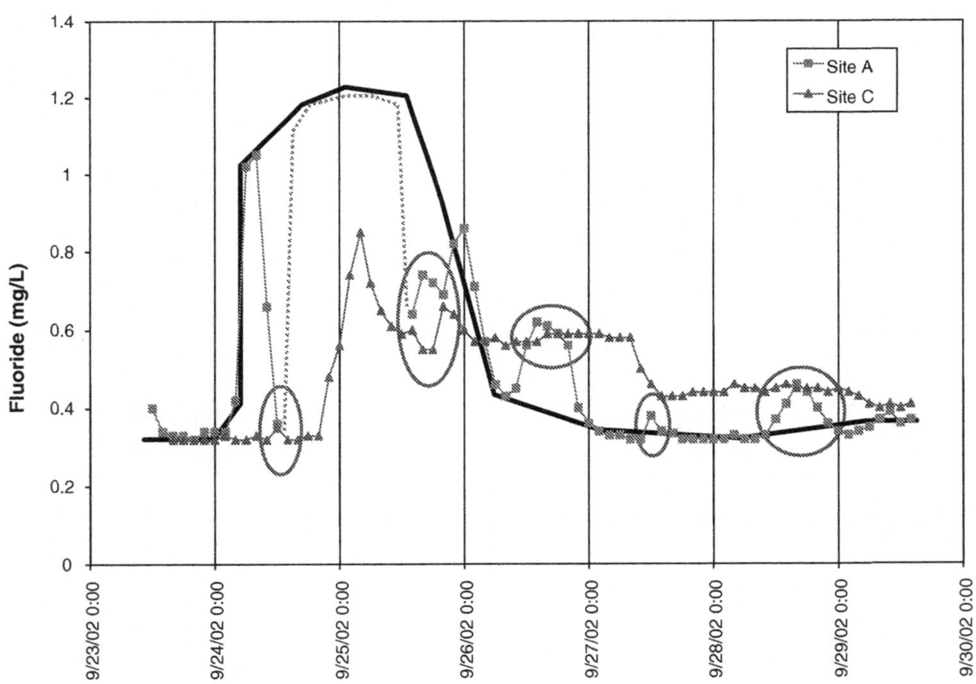


**Figure 6.8 Fluoride response curve showing influence of reservoir water releases**

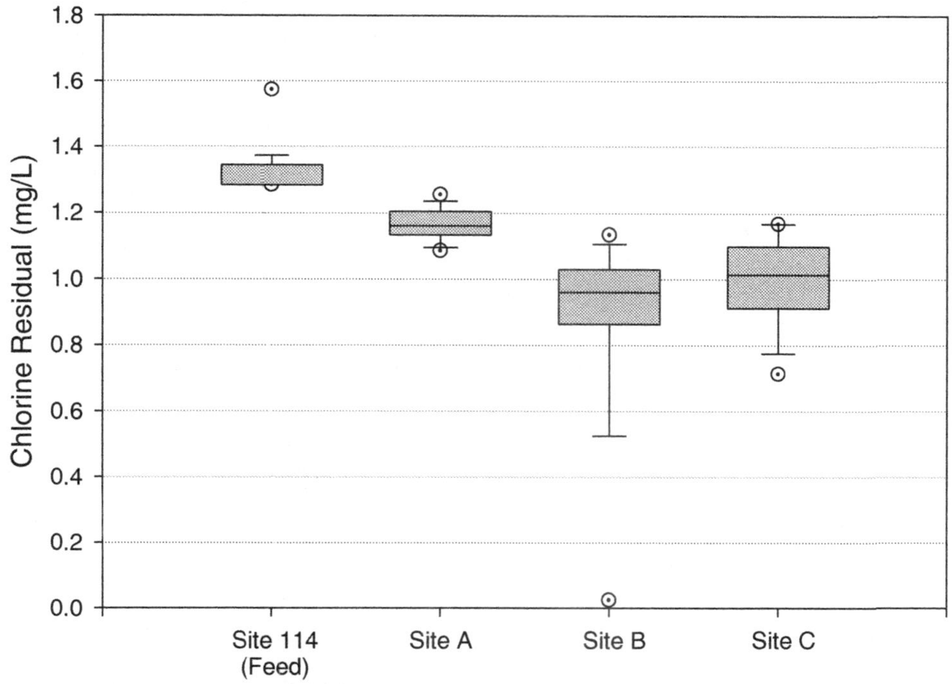


**Figure 6.9 Chlorine residual in Scottsdale's distribution system**

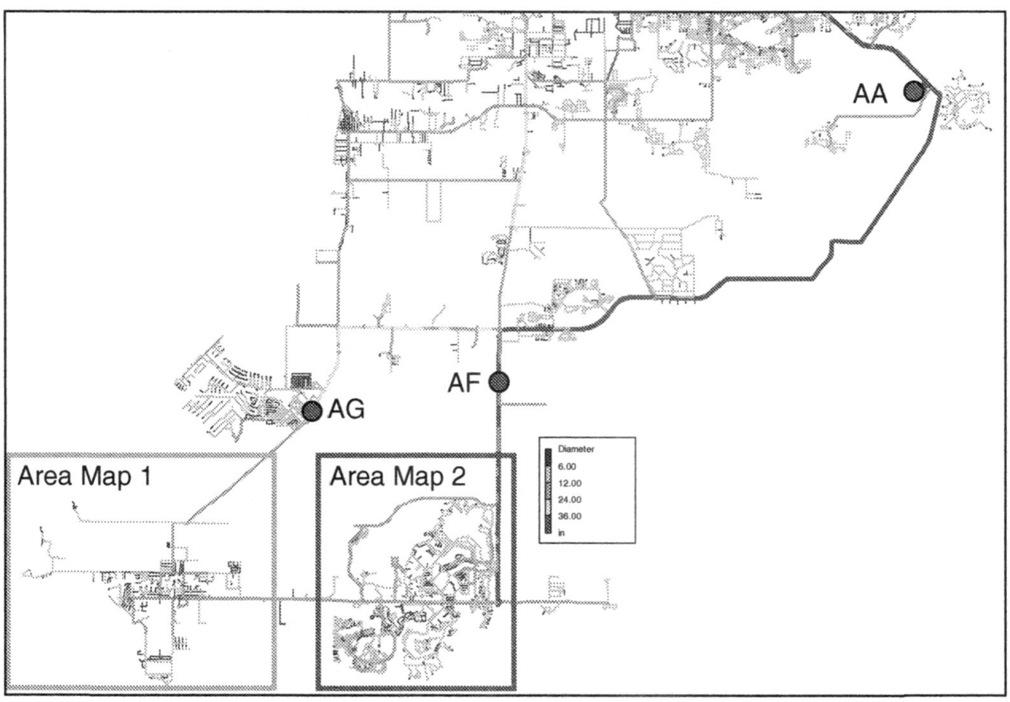


**Figure 6.10 Network study area. Red circles are water quality sampling locations. Location AA is the WTP, whereas AF and AG are transmission mains that constitute alternate flow paths to the south-western study endpoint.**

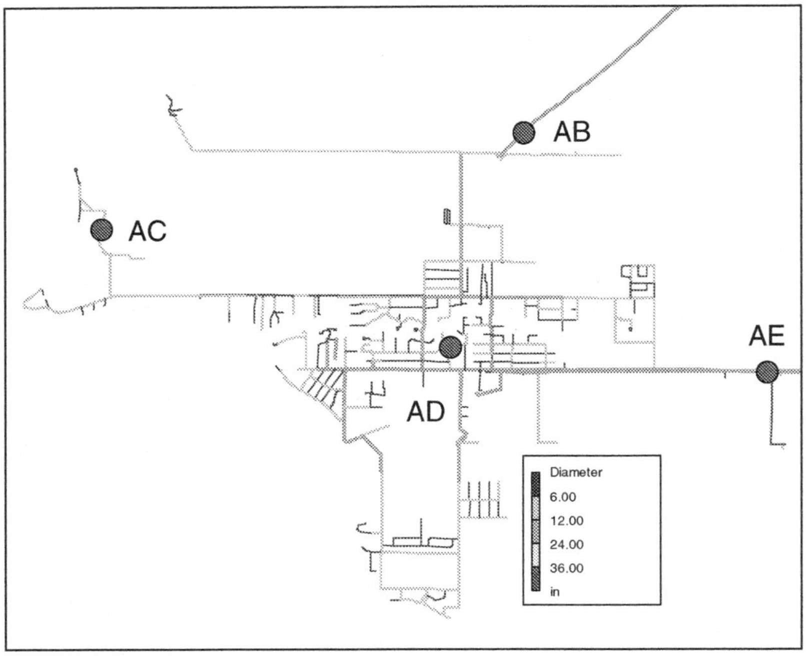


**Figure 6.11** Network subarea 1. Red circles are water quality sampling locations. Locations AB and AE are on the two flow paths feeding subarea 1; AC is on a dead end main that defines the western service area boundary; AD is in a highly looped residential section, on a hydraulic boundary between the northern and eastern flow paths that feed subarea 1.

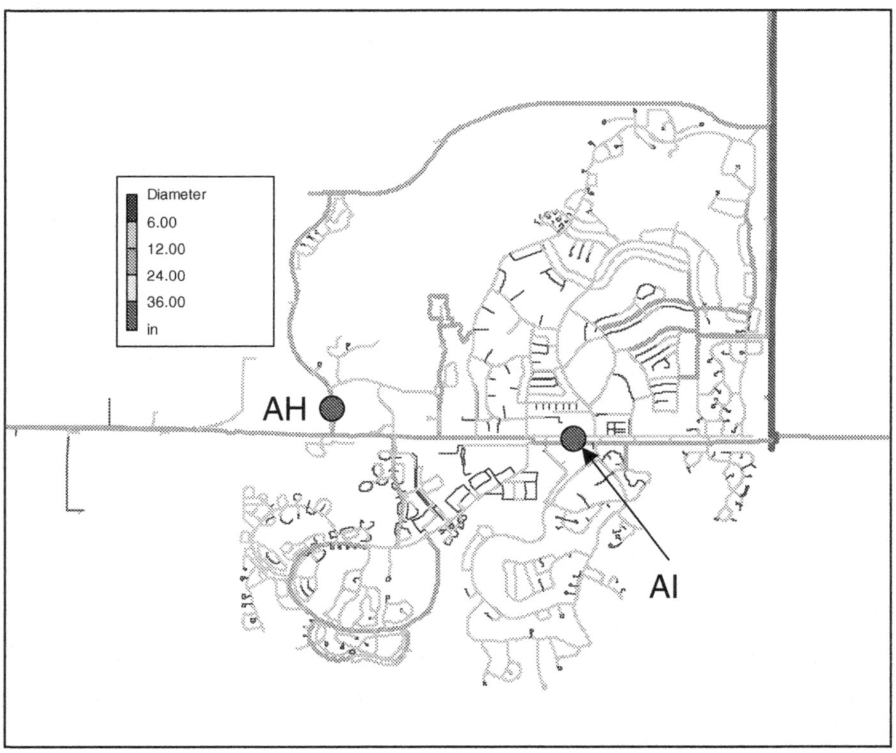


**Figure 6.12** Network subarea 2. Red circles are water quality sampling locations. Locations AH and AI are on principle flow paths through subarea 2 leading to subarea 1.

**Figure 6.13 Continuous conductivity and temperature analyzer and grab sample tap (typical)**

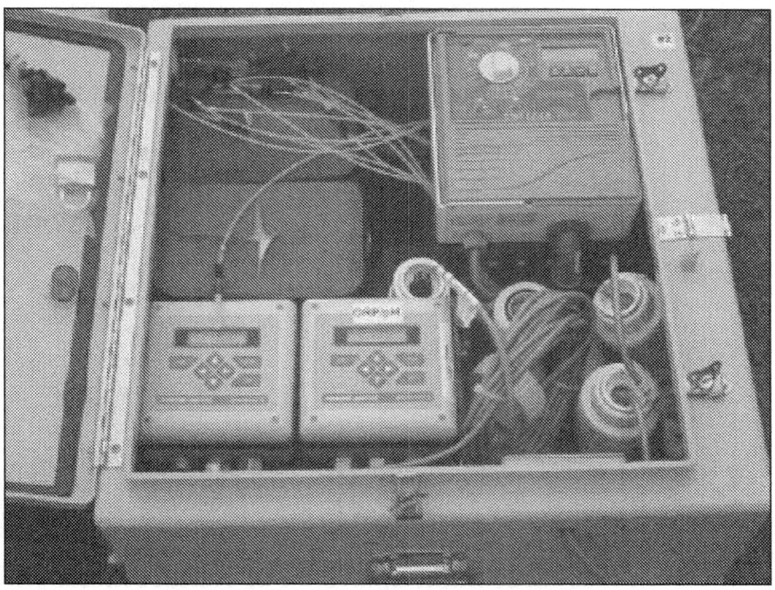

**Figure 6.14 Continuous chlorine and pH analyzer (typical)**

**Figure 6.15 Brine solution tank and injection pump**

**Figure 6.16 Brine injection into force main upstream of venturi meter (injection is on top of pipe, just after venturi meter flange)**

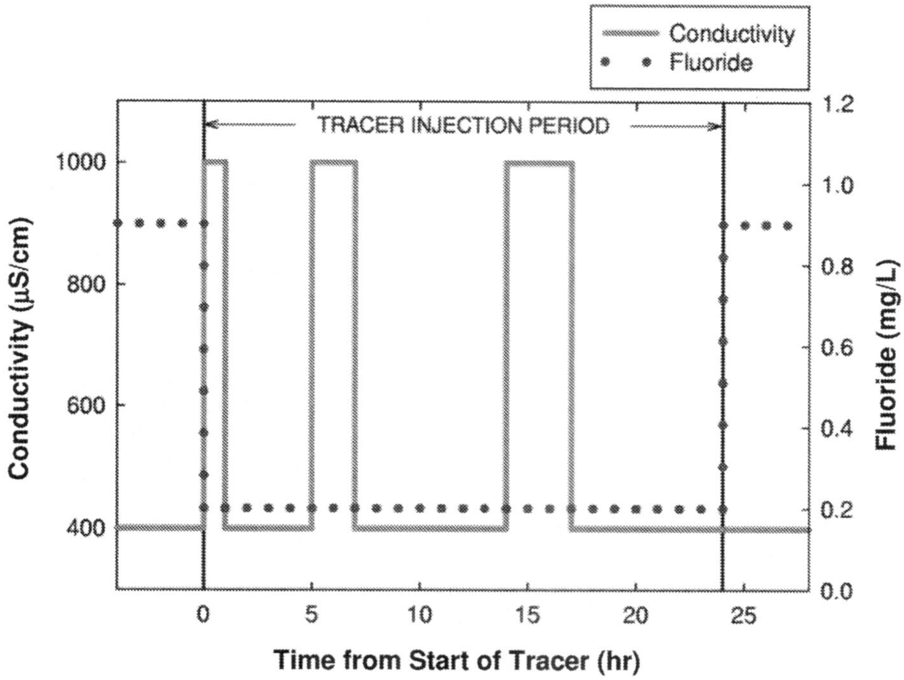


**Figure 6.17 Estimated conductivity and fluoride concentration leaving the treatment plant during the 24-hour injection period**

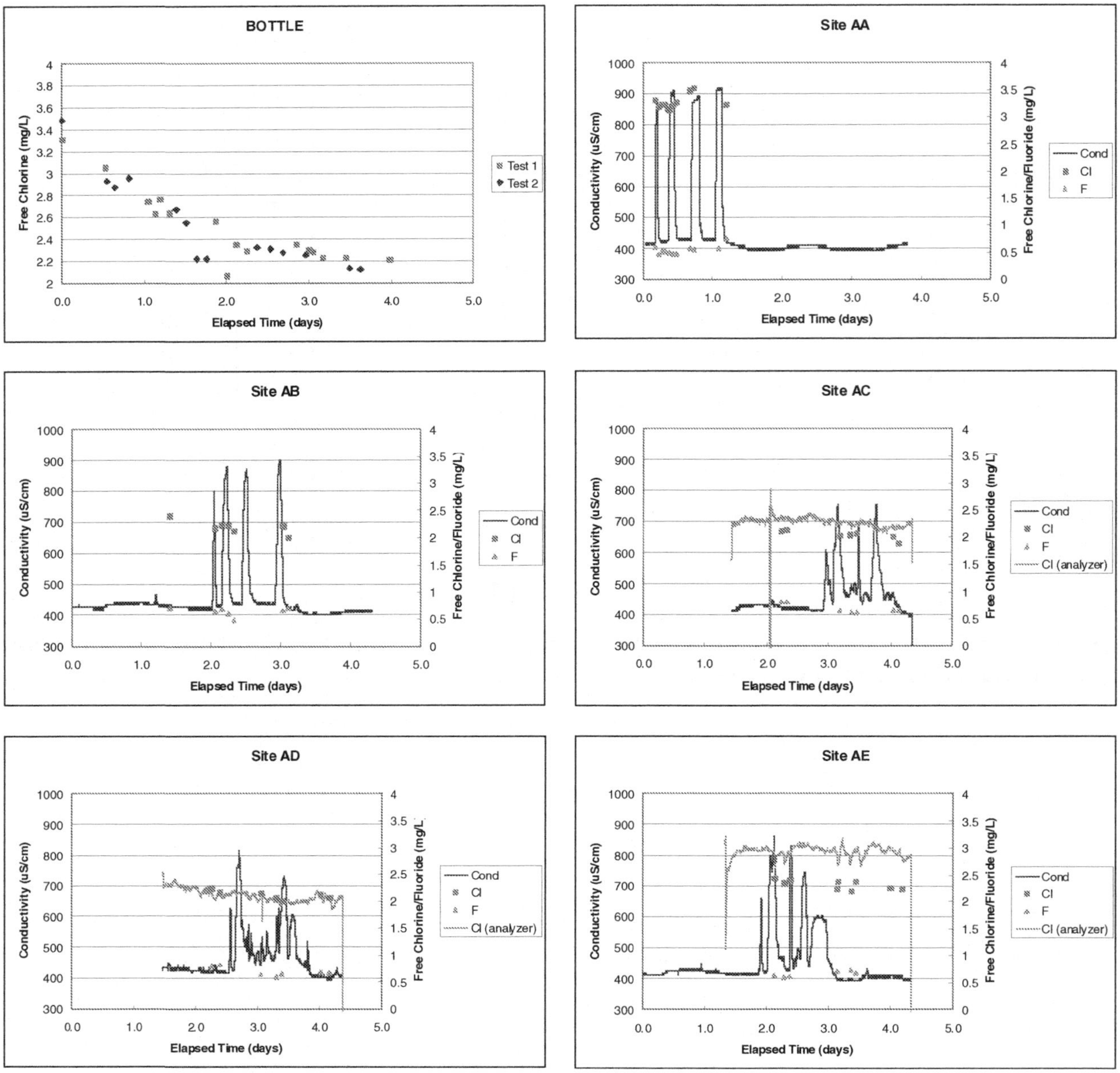


**Figure 6.18 Raw conductivity, and free chlorine and fluoride ion concentration, measured using continuous monitors and grab samples. Sample locations AA-AE, plus bottle test results.**

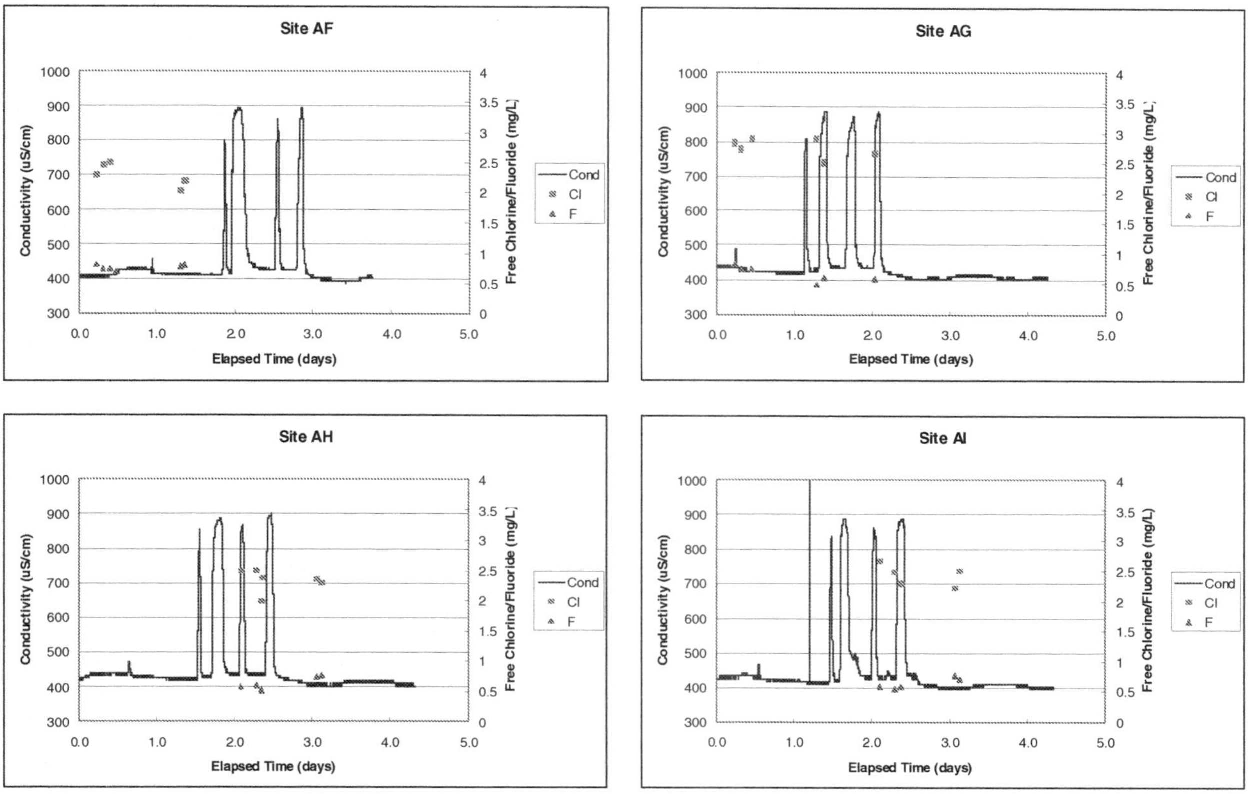


**Figure 6.19 Raw conductivity, and free chlorine and fluoride ion concentration, measured using continuous monitors and grab samples. Sample locations AF-AI.**

# CHAPTER 7
# CONCLUSIONS AND RECOMMENDATIONS TO UTILITIES

## FACTORS AFFECTING DISINFECTANT DECAY

### Effect of Advanced Treatment

The primary objective of this study was to determine the effect of advanced treatment on chlorine and chloramine residual decay in the distribution system. Residual disinfectant decay was measured at the bench-scale and pilot-scale. The bench-scale testing provided data indicating that treatment level affects residual decay for chlorine and chloramine in the bulk water. All of the advanced treatments were performed on water that had already been treated by enhanced coagulation. The following observations were made at the bench scale:

- EC slightly lowered the disinfectant demand as compared to raw water.
- UV irradiation did not affect disinfectant demand, suggesting that UV light did not significantly alter the water quality parameters that have an effect on disinfectant decay.
- GAC adsorption decreased the disinfectant demand, likely due to reduction in TOC.
- Ozone increased the demand, suggesting that when ozone reacts with organic matter, the byproducts of oxidation are either more reactive or are more numerous compared to the parameters before oxidation.
- Ozone/biofiltration decreased the demand, suggesting that biofiltration removes some of the reactive sites created during ozonation.

Pilot-scale testing was used to determine the effect of advanced treatment processes on the pipe-wall demand. One should keep in mind that these results were obtained for a single metallic pipe that has undergone appreciable corrosion, tubercle buildup, and biofilm colonization. They are, therefore, only indicative, and not generally conclusive, of what might be expected if treatment changes are made in systems containing pipes of a similar nature. The following observations were made based on pilot-scale evaluation in the pipe loop:

- The pipe wall demand rates for free chlorine added to ozonated or GAC treated water were about two times greater than those for conventional or RO treated water. This behavior may be related to the effect that changes in organic content have on iron complexation chemistry as well as the effect that changes in water conductivity have on the release of iron from corrosion reactions at the pipe wall.
- The pipe wall demand rates for chloraminated water showed no statistically significant variation with type of treatment received. This might reflect the possibility that wall reactions involving chloramines follow first-order kinetics more closely than those for free chlorine, whose kinetics are also apparently influenced by the rate at which reductants are made available at the pipe wall.

**Table 7.1**
**Summary of factors affecting chlorine decay**

| Factor | Effect on chlorine decay |
|---|---|
| Chlorine dose | As chlorine dose increased the relative demand also increased, resulting in a nearly identical 120-hour chlorine demand. |
| Temperature | Temperature has minimal effect during the first 3 hours of decay. After that, disinfectant residual decays faster at higher temperatures. |
| pH | Chlorine decay changes with pH; however, there is no definitive pattern, and the effect is specific to the water; the effect of changing pH in the middle of a decay study causes the curve to shift to the "new" pH curve. |
| Pipe sediments | As the amount of iron pipe sediment increased the decay rate also increased, resulting in lower chlorine residuals. Most of the decay (75-90 percent) occurred within the first 3 hours of the test. |
| Blending | Blending waters with different ages (and therefore chlorine residuals) results in a rechlorination of the "old" water." Blended water has more stable chlorine residual than either the new or the old water. |
| Rechlorination | The residual decay after rechlorination depends only on total chlorine dose, not the breakdown, i.e., a dose of 4 mg/L leaving the WTP is the same as 2 mg/L leaving the plant plus 2 mg/L at a booster station. |

### Effect of Additional Factors

In addition to treatment, several other factors were investigated to determine their effects on residual decay. The effects on chlorine and chloramine decay in the bulk phase are summarized in Table 7.1 and Table 7.2, respectively.

### Comparison of Wall and Bulk Reaction Rates

For the unlined ductile iron pipe used in the pilot-scale studies, chlorine decay was dominated by wall reactions. Rate constants for the wall reactions were 1 to 2 orders of magnitude higher than that for the bulk reactions for both free chlorine and chloramine. Wall reaction rate constants for free chlorine ranged from 1.1 to 4.6 days$^{-1}$, depending on chlorine concentration and treatment. These numbers are consistent with the rates measured by others. The wall rate constants for chloramines ranged only between 1.2 and 1.5 days$^{-1}$. Paired comparisons of wall rates for similar treatments showed the chloramine constants to be only 0.40 to 0.70 times the values of the free chlorine constants. Bulk rate constants were also lower for chloramine when compared to free chlorine.

## DEVELOPMENT OF WATER QUALITY MODELS

Another objective of this research was to develop models that can predict the bulk water chlorine and chloramine demand based on water quality parameters. The results of the bench-scale

**Table 7.2**
**Summary of factors affecting chloramine decay**

| Factor | Effect on chloramine decay |
|---|---|
| Chloramine dose | Relative decay kinetics for chloramine were independent of dose |
| Temperature | Temperature has minimal effect during the first 8 hours of decay. After that, disinfectant residual decays faster at higher temperatures. |
| pH | The effect of pH on chloramine decay was inconclusive. |
| Pipe sediments | Evaluation was not performed. |
| Blending | Blending water of different ages did not have an effect on chloramine residual because the chloramine concentration in water of different ages was similar. When waters with different residuals were blended, the resulting chloramine residual was the same as the weighted average of the two waters. |
| Rechloramination | The residual decay after rechloramination depends only on total chloramine dose, not the breakdown. |

studies generated several models that utilities will be able to use to predict chlorine demand after changes to treatment.

- **Chlorine Demand Model:** Chlorine demand at 1, 24, or 120 hours can be predicted as a function of TOC, UVA, and $Cl_2$ dose at chlorine dose ratios between 0.75 and 2.5 mg/L C12:mg/L TOC. The equation of this model is the following form:

$$CD_t = \chi\,(C_0) + \tau\,(TOC) + \upsilon\,(UVA) \tag{7.1}$$

where $CD_t$ = chlorine demand at time t = 1, 24, or 120 hours (mg/L)
$C_0$ = chlorine dose (mg/L)
TOC = total organic carbon concentration (mg/L)
UVA = ultraviolet absorbance at 254 nanometers ($cm^{-1}$)
$\chi, \tau, \upsilon$ = regression coefficients, $\chi,\tau$ (dimensionless), $\upsilon$ (cm·mg/L)

The regression coefficients for time = 1, 24, and 120 hours are presented in Table 3.10.

- **Chlorine Decay Curve Model for Change in Chlorine Dose:** The chlorine decay curve for new chlorine dose can be predicted using Equation 7.2 given a known decay curve for a specific water and one long-term demand point ($CD_{24}$ or $CD_{120}$). The demand point can either be determined experimentally or from the Chlorine Demand Model

$$C_t = C_0 - \left(\frac{CD_t^*}{CD_{120}^*}\right) \cdot CD_{120} \tag{7.2}$$

where $C_t$ = Chlorine residual at time, t, at new chlorine dose (mg/L)
$C_0$ = New chlorine dose (mg/L)
$CD_t^*$ = Chlorine demand at time, t, for old chlorine dose (mg/L)
$CD_{120}^*$ = Chlorine demand at 120 hours for old chlorine dose (mg/L)
$CD_{120}$ = Chlorine demand at 120 hours for new chlorine dose (mg/L)

- **Chlorine Decay Curve Model for Change in Treatment:** Chlorine decay curves for new treatment processes can be predicted using Equation 7.3 given a known chlorine decay curve and a 1-hour chlorine residual. The 1-hour chlorine residual can be determined experimentally or using Equation 7.4.

$$C_t = C_1 - \left(\frac{C_t^*}{C_1^*}\right) \quad (7.3)$$

where $C_t$ = Chlorine residual at time t after new treatment (mg/L)
$C_1$ = Chlorine residual at 1 hour after new treatment (mg/L)
$C_t^*$ = Chlorine residual at time t before new treatment (mg/L)
$C_1^*$ = Chlorine residual at 1 hour before new treatment (mg/L)

$$C_1 = C_0 - CD_1 \quad (7.4)$$

where $C_1$ = chlorine residual in water at 1 hour after new treatment (mg/L)
$C_0$ = initial chlorine concentration (mg/L)
$CD_1$ = 1-hour chlorine demand predicted via Equation 7.1 and Table 3.10

Due to limited trends and correlations in the chloramine data, only one model was developed: the chloramine demand predictions at 1, 5, or 20 days as a function of TOC, UVA and $Cl_2$ dose (Equation 7.5). The model is not as robust as the chlorine model, but it is still a good predictive tool.

$$CLMD_t = \chi\,(C_0) + \tau\,(TOC) + \upsilon\,(UVA) \quad (7.5)$$

where $CLMD_t$ = chlorine demand at time, t = 1, 5, or 20 days (mg/L)
$C_0$ = chloramine dose (mg/L)
TOC = total organic carbon concentration (mg/L)
UVA = ultraviolet absorbance at 254 nanometers ($cm^{-1}$)
$\chi$, $\tau$, $\upsilon$ = regression coefficients, $\chi$,$\tau$ (dimensionless), $\upsilon$ (cm·mg/L)

The regression coefficients for time = 1, 5, and 20 days are presented in Table 3.12.

## EPANET EXTENSIONS

The multi-species water quality analysis and Network Path Analysis extensions to EPANET that were developed for this project will be a valuable addition to the modeling software currently available to utilities. These tools are a direct modifications to the *C* code that implements the *Epanet2 Programmer's Toolkit* (Rossman 1999). This toolkit is a set of numerical algorithms that implement the *Epanet2* hydraulic and water quality network algorithms and a set of function definitions that allow access to them from programs written in the *C* programming language.

The updates to the algorithms allow for multi-species chlorine modeling and also incorporate an innovated method of identifying hydraulic flow paths between points. The multi-species model allows for the model to track the chemical reactions that take place in the distribution system. For instance, the reactions between chlorine and NOM to form DBPs or chlorine and ammonia to form chloramine. The network path analysis incorporates the idea that not all water at a given point arrived there via the same path. This software will make the modeling package more robust and allow utilities to predict chlorine residual at any point in their distribution system with greater confidence. The *Epanet2* desktop application is built using the *Programmer's Toolkit*, and it has been used worldwide in a host of applications. Thus, by implementing the multi-species modifications to the toolkit code, the multi-species capability becomes accessible to a similar wide range of applications.

## FIELD-SCALE TESTING

Studying chlorine and chloramine residual decay on the field-scale is site-specific. Some general conclusions from the studies performed for this project are as follows:

- Tracer testing can be used to pinpoint water age.
- Tracer testing is a viable method for determining various pathways from the source to the sample location and for determining the influence of multiple sources.
- Fluoride and conductivity are two materials that are conservative with respect to concentration and are also relatively easy to measure, making them good selections for performing tracer studies.

With respect to the study performed at the Southeastern United States Utility, some more specific conclusions can be made:

- A conductivity pulse of one to several hours can successfully be traced through the distribution system. Longer tracer feed is not necessary.
- The increased conductivity during the tracer study does not need to be double the background concentration. Instead, it only needs to be high enough to see a measurable difference. The benefits of this are that utilities with a naturally high conductivity do not need to be concerned about violating maximum contaminant levels and the amount of brine solution needed for injection can be reduced.
- The accuracy of conductivity monitors designed for in-process use performed nearly flawlessly. The monitors used in this specific study were able to operate on batteries

for approximately 1 week, which allowed for the monitors to be set up on a different day from beginning the tracer injection.

- The most challenging part of performing the tracer study was setting up the tracer injection, especially in regards to using flow pacing to control the concentration after injection.

# APPENDIX A
# CHLORINE DECAY RAW DATA

**Table A.1**
**Measured experimental conditions table**

| | | |
|---|---|---|
| pH | Base case | 8.0 ± 0.1 |
| | Low pH | 5.5 ± 0.3 |
| | High pH | 10.5 ± 0.2 |
| Temperature | Base case | 20°C ± 2°C |
| | Low temp | 4°C ± 2°C |
| | High temp | 30°C ± 2°C |

Only non-base case conditions noted in tables

**Table A.2**
**Betasso treatment studies (1.5:1 $Cl_2$:TOC mg/mg)**

| Raw | | EC | | UV | | GAC | | $O_3$/BF | |
|---|---|---|---|---|---|---|---|---|---|
| Time (hr) | $Cl_2$ residual (mg/L) | Time (hr) | $Cl_2$ residual (mg/L) | Time (hr) | $Cl_2$ residual (mg/L) | Time (hr) | $Cl_2$ residual (mg/L) | Time (hr) | $Cl_2$ residual (mg/L) |
| Dose | 4.46 | Dose | 2.32 | Dose | 2.31 | Dose | 1.73 | Dose | 1.68 |
| 0.048 | 3.7 | 0.063 | 1.90 | 0.05 | 1.70 | 0.060 | 1.37 | 0.067 | 1.50 |
| 0.094 | 3.55 | 0.115 | 1.58 | 0.11 | 1.59 | 0.109 | 1.31 | 0.129 | 1.45 |
| 0.176 | 3.4 | 0.170 | 1.55 | 0.18 | 1.51 | 0.191 | 1.26 | 0.217 | 1.41 |
| 0.232 | 3.32 | 0.258 | 1.45 | 0.25 | 1.42 | 0.286 | 1.20 | 0.256 | 1.41 |
| 0.496 | 3.2 | 0.499 | 1.35 | 0.53 | 1.28 | 0.482 | 1.10 | 0.509 | 1.25 |
| 0.742 | 3.05 | 0.755 | 1.28 | 0.80 | 1.25 | 0.750 | 1.03 | 0.763 | 1.28 |
| 1.003 | 3 | 1.036 | 1.20 | 1.03 | 1.25 | 1.011 | 1.02 | 1.081 | 1.295 |
| 1.053 | 2.95 | 1.108 | 1.30 | 1.11 | 1.20 | 1.056 | 1.02 | 1.056 | 1.30 |
| 1.493 | 2.8 | 1.515 | 1.17 | 1.62 | 1.18 | 1.556 | 0.98 | 1.106 | 1.29 |
| 1.993 | 2.75 | 2.062 | 1.10 | 1.94 | 1.19 | 1.993 | 0.95 | 1.494 | 1.24 |
| 2.488 | 2.67 | 2.100 | 1.18 | 1.99 | 1.10 | 2.494 | 0.95 | 1.998 | 1.24 |
| 2.554 | 2.65 | 2.472 | 1.15 | 2.68 | 1.00 | 2.588 | 0.91 | 2.694 | 1.22 |
| 3.086 | 2.57 | 3.286 | 1.15 | 3.00 | 1.08 | 3.037 | 0.91 | 3.417 | 1.11 |
| 5.045 | 2.31 | 4.965 | 1.05 | 5.07 | 0.97 | 5.036 | 0.82 | 5.076 | 1.13 |
| 7.989 | 2.15 | 5.026 | 1.05 | 5.19 | 0.99 | 5.083 | 0.85 | 5.119 | 1.15 |
| 24.19 | 1.55 | 7.769 | 0.97 | 8.22 | 0.93 | 9.130 | 0.68 | 9.390 | 1.10 |
| 24.44 | 1.52 | 23.96 | 0.70 | 24.30 | 0.60 | 9.163 | 0.77 | 9.433 | 1.05 |
| 52.01 | 1.07 | 24.09 | 0.75 | 24.40 | 0.62 | 24.42 | 0.61 | 24.57 | 0.73 |
| 71.95 | 0.76 | 69.55 | 0.40 | 48.33 | 0.44 | 24.55 | 0.58 | 24.62 | 0.90 |
| 121.4 | 0.4 | 69.58 | 0.45 | 76.37 | 0.225 | 57.51 | 0.41 | 47.95 | 0.68 |
| | | 92.80 | 0.35 | 123.0 | 0.05 | 77.58 | 0.35 | 73.39 | 0.55 |
| | | 92.86 | 0.37 | 123.0 | 0.05 | 120.3 | 0.16 | 123.2 | 0.42 |
| | | 122.2 | 0.15 | | | | | 123.2 | 0.43 |
| | | 122.2 | 0.24 | | | | | | |

**Table A.3**
**Betasso treatment studies (2.5:1 $Cl_2$:TOC mg/mg)**

| Raw | | EC | | GAC | | UV | | $O_3$/BF | |
|---|---|---|---|---|---|---|---|---|---|
| Time (hr) | $Cl_2$ residual (mg/L) | Time (hr) | $Cl_2$ residual (mg/L) | Time (hr) | $Cl_2$ residual (mg/L) | Time (hr) | $Cl_2$ residual (mg/L) | Time (hr) | $Cl_2$ residual (mg/L) |
| Dose | 7.35 | Dose | 3.81 | Dose | 3.79 | Dose | 2.86 | Dose | 2.73 |
| 0.049 | 6.07 | 0.057 | 3.00 | 0.056 | 3.10 | 0.061 | 2.47 | 0.047 | 2.49 |
| 0.095 | 5.97 | 0.100 | 2.90 | 0.115 | 2.95 | 0.108 | 2.45 | 0.097 | 2.42 |
| 0.167 | 5.85 | 0.169 | 2.80 | 0.180 | 2.88 | 0.173 | 2.35 | 0.188 | 2.38 |
| 0.264 | 5.73 | 0.256 | 2.70 | 0.310 | 2.85 | 0.255 | 2.31 | 0.257 | 2.38 |
| 0.614 | 5.73 | 0.753 | 2.60 | 0.513 | 2.70 | 0.522 | 2.15 | 0.546 | 2.25 |
| 0.788 | 5.50 | 0.986 | 2.50 | 0.805 | 2.64 | 0.764 | 2.15 | 0.745 | 2.26 |
| 1.017 | 5.40 | 1.035 | 2.60 | 0.979 | 2.59 | 1.024 | 2.10 | 1.020 | 2.24 |
| 1.581 | 5.08 | 1.503 | 2.50 | 1.551 | 2.56 | 1.070 | 2.11 | 1.083 | 2.21 |
| 1.978 | 5.17 | 1.989 | 2.55 | 1.969 | 2.50 | 1.509 | 2.05 | 1.448 | 2.22 |
| 2.457 | 5.10 | 2.417 | 2.40 | 2.729 | 2.45 | 1.986 | 2.04 | 2.023 | 2.15 |
| 2.496 | 5.15 | 2.483 | 2.40 | 3.261 | 2.55 | 2.817 | 1.98 | 2.830 | 2.17 |
| 2.983 | 4.95 | 3.000 | 2.40 | 3.310 | 2.45 | 3.298 | 1.94 | 3.135 | 2.09 |
| 4.990 | 4.77 | 4.994 | 2.35 | 5.018 | 2.40 | 5.930 | 1.95 | 5.004 | 2.07 |
| 8.069 | 4.37 | 8.017 | 2.00 | 5.179 | 2.35 | 5.004 | 1.93 | 5.076 | 2.06 |
| 24.23 | 3.73 | 8.100 | 2.00 | 8.119 | 2.26 | 8.064 | 1.84 | 8.901 | 2.03 |
| 24.30 | 3.75 | 23.96 | 1.85 | 26.8 | 1.90 | 8.120 | 1.85 | 8.939 | 2.03 |
| 49.29 | 3.15 | 24.01 | 1.90 | 26.8 | 1.90 | 24.02 | 1.56 | 24.06 | 1.81 |
| 72.12 | 2.83 | 48.17 | 1.43 | 47.02 | 1.60 | 24.16 | 1.60 | 24.11 | 1.72 |
| 118.7 | 2.25 | 72.02 | 1.35 | 73.05 | 1.35 | 56.40 | 1.32 | 47.294 | 1.45 |
| | | 119.8 | 1.07 | 121.7 | 0.85 | 76.51 | 1.20 | 72.795 | 1.38 |
| | | | | 121.7 | 1.01 | 119.4 | 1.05 | 122.6 | 1.15 |
| | | | | | | 119.4 | 1.00 | 122.6 | 1.16 |

**Table A.4**
**Betasso EC temperature studies (2.5:1 $Cl_2$:TOC)**

| 4°C | | 30°C | |
|---|---|---|---|
| Time (hr) | $Cl_2$ residual (mg/L) | Time (hr) | $Cl_2$ residual (mg/L) |
| Dose | 3.82 | 0.000 | 3.80 |
| 0.068 | 2.95 | 0.058 | 2.91 |
| 0.101 | 2.90 | 0.103 | 2.82 |
| 0.172 | 2.77 | 0.161 | 2.76 |
| 0.289 | 2.75 | 0.261 | 2.70 |
| 0.516 | 2.65 | 0.493 | 2.54 |
| 0.726 | 2.65 | 0.758 | 2.50 |
| 0.991 | 2.60 | 0.976 | 2.49 |
| 1.033 | 2.55 | 1.021 | 2.46 |
| 1.478 | 2.53 | 1.528 | 2.45 |
| 2.008 | 2.57 | 1.994 | 2.33 |
| 2.515 | 2.50 | 2.042 | 2.34 |
| 2.992 | 2.50 | 2.494 | 2.30 |
| 3.103 | 2.40 | 2.994 | 2.20 |
| 5.226 | 2.48 | 3.051 | 2.22 |
| 8.040 | 2.39 | 4.979 | 2.05 |
| 24.00 | 2.35 | 7.507 | 1.93 |
| 24.02 | 2.71 | 23.85 | 1.37 |
| 48.52 | 2.00 | 47.95 | 1.06 |
| 73.55 | 1.75 | 71.93 | 0.80 |
| 120.0 | 1.57 | 71.97 | 0.85 |
| | | 121.3 | 0.41 |

**Table A.5**
**Betasso EC pH studies (1.5:1 $Cl_2$:TOC)**

| pH 5.5 | | pH 10.5 | | Changing pH (5.5 to 8.0) | |
|---|---|---|---|---|---|
| Time (hr) | $Cl_2$ residual (mg/L) | Time (hr) | $Cl_2$ residual (mg/L) | Time (hr) | $Cl_2$ residual (mg/L) |
| Dose | 2.30 | Dose | 2.30 | Dose | 2.31 |
| 0.050 | 1.91 | 0.065 | 1.84 | 0.064 | 1.88 |
| 0.079 | 1.93 | 0.103 | 1.70 | 0.094 | 1.80 |
| 0.174 | 1.88 | 0.159 | 1.68 | 0.150 | 1.75 |
| 0.245 | 1.88 | 0.246 | 1.56 | 0.258 | 1.64 |
| 0.533 | 1.70 | 0.517 | 1.45 | 0.449 | 1.57 |
| 0.746 | 1.72 | 0.808 | 1.40 | 0.983 | 1.45 |
| 0.974 | 1.70 | 0.991 | 1.40 | 1.033 | 1.53 |
| 1.031 | 1.67 | 1.030 | 1.39 | 1.734 | 1.39 |
| 1.461 | 1.70 | 1.586 | 1.28 | 2.566 | 1.31 |
| 1.973 | 1.60 | 1.963 | 1.27 | 2.610 | 1.36 |
| 2.495 | 1.53 | 2.493 | 1.29 | 2.957 | 1.34 |
| 2.544 | 1.55 | 2.526 | 1.32 | 5.061 | 1.22 |
| 3.028 | 1.52 | 3.006 | 1.24 | 5.221 | 1.20 |
| 4.976 | 1.39 | 4.581 | 1.15 | 5.247 | 1.21 |
| 7.383 | 1.35 | 8.467 | 1.00 | 6.266 | 1.11 |
| 29.86 | 0.95 | 29.86 | 0.95 | 6.340 | 1.13 |
| 29.86 | 0.85 | 29.86 | 0.92 | 7.091 | 1.08 |
| 45.08 | 0.52 | 46.90 | 0.82 | 8.133 | 1.06 |
| 45.88 | 0.50 | 46.98 | 0.80 | 8.189 | 1.03 |
| 70.63 | 0.32 | 71.62 | 0.61 | 29.07 | 0.75 |
| 119.8 | 0.16 | 119.8 | 0.60 | 29.13 | 0.75 |
| 119.8 | 0.18 | 119.8 | 0.59 | 51.80 | 0.58 |
| | | | | 77.79 | 0.48 |
| | | | | 126.9 | 0.48 |
| | | | | 126.9 | 0.47 |

**Table A.6**
**Betasso blending studies**

| 2-hour / 24-hour | | 2-hour / 5-day | |
|---|---|---|---|
| Time (hr) | $Cl_2$ Residual (mg/L) | Time (hr) | $Cl_2$ Residual (mg/L) |
| 0 | Old | 0 | Old |
| Dose | 1.73 | Dose | 1.73 |
| 0.055 | 1.32 | 0.938 | 1.43 |
| 1.127 | 0.96 | 2.808 | 1.10 |
| 2.506 | 0.95 | 23.89 | 0.68 |
| 23.60 | 0.72 | 51.21 | 0.56 |
| 23.35 | New | 71.65 | 0.50 |
| Dose | 1.73 | 118.78 | 0.37 |
| 23.41 | 1.27 | 118.99 | New |
| 24.25 | 1.04 | 118.99 | 1.73 |
| 25.38 | 0.98 | 119.03 | 1.41 |
| 25.46 | Blend | 120.20 | 1.08 |
| 25.88 | 0.80 | 121.23 | 1.02 |
| 26.19 | 0.80 | 121.32 | Blend |
| 28.38 | 0.70 | 121.48 | 0.72 |
| 30.73 | 0.77 | 122.09 | 0.69 |
| 30.77 | 0.72 | 124.40 | 0.60 |
| 48.45 | 0.58 | 124.43 | 0.62 |
| 48.57 | 0.61 | 126.89 | 0.63 |
| 71.15 | 0.52 | 126.93 | 0.59 |
| 97.54 | 0.42 | 143.79 | 0.50 |
| 145.0 | 0.30 | 145.00 | 0.50 |
| 145.0 | 0.31 | 166.88 | 0.41 |
| | | 189.59 | 0.33 |

**Table A.7**
**Betasso rechlorination studies**

| 5-hour | | 24-hour | |
|---|---|---|---|
| Time (hr) | $Cl_2$ Residual (mg/L) | Time (hr) | $Cl_2$ Residual (mg/L) |
| Dose | 2.25 | Dose | 2.25 |
| 0.071 | 1.57 | 0.071 | 1.57 |
| 0.108 | 1.48 | 0.108 | 1.48 |
| 0.992 | 1.17 | 0.992 | 1.17 |
| 4.457 | 0.98 | 4.457 | 0.98 |
| 26.19 | 0.67 | 8.000 | 0.90 |
| $ReCl_2$ | 1.52 | 26.19 | 0.67 |
| 4.691 | 2.37 | $ReCl_2$ | 1.52 |
| 4.754 | 2.37 | 26.52 | 2.17 |
| 8.124 | 2.21 | 26.57 | 2.20 |
| 8.187 | 2.27 | 48.09 | 1.87 |
| 26.26 | 1.90 | 48.20 | 1.89 |
| 26.31 | 1.86 | 79.27 | 1.65 |
| 47.96 | 1.65 | 79.37 | 1.50 |
| 79.04 | 1.46 | 145.9 | 1.21 |
| 79.16 | 1.36 | 145.9 | 1.28 |
| 145.8 | 1.25 | | |
| 145.8 | 1.24 | | |

**Table A.8**
**Betasso EC duplicate study (2.5:1 $Cl_2$:TOC)**

| Time (hr) | $Cl_2$ residual (mg/L) | Time (hr) | $Cl_2$ residual (mg/L) |
|---|---|---|---|
| Dose | 3.75 | 73.41 | 1.47 |
| 0.061 | 2.92 | 73.53 | 1.57 |
| 0.992 | 2.55 | 100.0 | 1.42 |
| 5.017 | 2.32 | 100.0 | 1.42 |
| 24.28 | 1.94 | 119.3 | 1.37 |
| 24.40 | 1.93 | 119.3 | 1.34 |

**Table A.9**
**NPNV treatment studies (1.5:1 $Cl_2$:TOC)**

| Time (hr) | $Cl_2$ residual (mg/L) | Time (hr) | $Cl_2$ residual (mg/L) | Time (hr) | $Cl_2$ residual (mg/L) |
|---|---|---|---|---|---|
| Raw | | EC | | UV | |
| Dose | 8.25 | Dose | 5.36 | Dose | 5.25 |
| 0.046 | 6.25 | 0.043 | 4.75 | 0.049 | 4.80 |
| 0.118 | 6.03 | 0.130 | 4.57 | 0.113 | 4.62 |
| 0.206 | 5.84 | 0.212 | 4.42 | 0.164 | 4.57 |
| 0.292 | 5.65 | 0.264 | 4.27 | 0.310 | 4.47 |
| 0.529 | 5.47 | 0.507 | 4.07 | 0.607 | 4.22 |
| 0.721 | 5.30 | 0.808 | 4.07 | 0.932 | 4.08 |
| 0.984 | 5.11 | 1.110 | 3.84 | 1.474 | 4.17 |
| 1.527 | 4.85 | 1.511 | 3.97 | 1.969 | 3.93 |
| 2.174 | 4.57 | 2.092 | 3.75 | 2.473 | 3.82 |
| 2.272 | 4.55 | 2.506 | 3.74 | 3.168 | 3.72 |
| 2.549 | 4.41 | 3.334 | 3.59 | 3.835 | 3.79 |
| 3.073 | 4.42 | 5.467 | 3.31 | 5.868 | 3.55 |
| 5.733 | 3.92 | 8.617 | 3.17 | 24.01 | 2.60 |
| 8.382 | 3.61 | 24.46 | 2.51 | 24.06 | 2.66 |
| 24.12 | 2.54 | 24.53 | 2.47 | 50.70 | 2.10 |
| 48.18 | 1.71 | 48.34 | 1.91 | 70.17 | 1.77 |
| 72.01 | 1.32 | 73.83 | 1.52 | 123.5 | 1.20 |
| 72.09 | 1.13 | 120.9 | 0.50 | 123.5 | 1.25 |
| 117.7 | 0.50 | | | | |
| GAC | | $O_3$ | | $O_3$/BF | |
| Dose | 3.86 | Dose | 4.25 | Dose | 3.17 |
| 0.063 | 3.55 | 0.044 | 3.78 | 0.046 | 2.83 |
| 0.101 | 3.55 | 0.094 | 3.70 | 0.092 | 2.80 |
| 0.148 | 3.49 | 0.155 | 3.58 | 0.189 | 2.73 |
| 0.280 | 3.39 | 0.240 | 3.45 | 0.279 | 2.63 |
| 0.555 | 3.27 | 0.533 | 3.25 | 0.506 | 2.53 |
| 0.979 | 3.10 | 0.723 | 3.16 | 0.758 | 2.55 |
| 1.483 | 3.10 | 1.066 | 3.09 | 1.001 | 2.48 |
| 2.038 | 3.00 | 1.563 | 2.95 | 1.058 | 2.47 |
| 2.076 | 2.94 | 2.128 | 2.84 | 1.475 | 2.44 |
| 2.541 | 2.90 | 2.169 | 2.86 | 2.485 | 2.32 |
| 2.965 | 2.87 | 2.572 | 2.80 | 2.547 | 2.32 |
| 5.257 | 2.71 | 2.993 | 2.68 | 3.107 | 2.29 |
| 25.52 | 1.92 | 3.048 | 2.68 | 4.405 | 2.18 |
| 25.56 | 1.99 | 23.19 | 1.58 | 4.975 | 2.14 |
| 50.99 | 1.68 | 23.24 | 1.59 | 7.732 | 2.02 |
| 70.31 | 1.49 | 4.901 | 2.50 | 24.08 | 1.61 |
| 122.4 | 1.10 | 12.20 | 2.00 | 24.14 | 1.62 |
| 122.4 | 1.05 | 47.45 | 0.75 | 49.66 | 1.28 |
| | | 47.50 | 0.90 | 72.00 | 1.15 |
| | | 70.90 | 0.49 | 119.10 | 0.78 |
| | | 119.7 | 0.13 | 119.10 | 0.80 |

**Table A.10**
**NPNV treatment studies (0.75:1 $Cl_2$:TOC)**

| EC | | $O_3$ only | | $O_3$/BF | |
|---|---|---|---|---|---|
| Time (hr) | $Cl_2$ residual (mg/L) | Time (hr) | $Cl_2$ residual (mg/L) | Time (hr) | $Cl_2$ residual (mg/L) |
| Dose | 2.67 | Dose | 2.14 | Dose | 1.59 |
| 0.079 | 2.19 | 0.048 | 1.69 | 0.046 | 1.35 |
| 0.237 | 2.01 | 0.088 | 1.65 | 0.123 | 1.23 |
| 0.506 | 1.82 | 0.136 | 1.65 | 0.225 | 1.27 |
| 0.988 | 1.69 | 0.234 | 1.51 | 0.274 | 1.23 |
| 3.051 | 1.40 | 0.453 | 1.34 | 0.513 | 1.14 |
| 3.108 | 1.46 | 0.733 | 1.22 | 0.835 | 1.04 |
| 8.071 | 1.15 | 1.031 | 1.14 | 0.998 | 1.06 |
| 25.84 | 0.65 | 1.521 | 1.06 | 1.483 | 1.02 |
| 53.73 | 0.30 | 2.051 | 0.95 | 2.011 | 0.94 |
| 53.79 | 0.37 | 2.605 | 0.90 | 2.926 | 0.88 |
| 72.26 | 0.16 | 3.032 | 0.75 | 3.016 | 0.85 |
| | | 3.066 | 0.82 | 3.809 | 0.84 |
| | | 5.025 | 0.70 | 5.107 | 0.80 |
| | | 5.095 | 0.70 | 7.748 | 0.69 |
| | | 12.70 | 0.30 | 7.983 | 0.68 |
| | | 23.68 | 0.06 | 24.08 | 0.37 |
| | | 47.93 | 0.00 | 24.13 | 0.37 |
| | | | | 49.07 | 0.13 |
| | | | | 49.13 | 0.14 |
| | | | | 71.99 | 0.05 |
| | | | | 119.6 | 0.00 |

**Table A.11**
**NPNV treatment studies (2.5:1 $Cl_2$:TOC)**

| Raw | | EC | | GAC | |
|---|---|---|---|---|---|
| Time (hr) | $Cl_2$ residual (mg/L) | Time (hr) | $Cl_2$ residual (mg/L) | Time (hr) | $Cl_2$ residual (mg/L) |
| Dose | 13.84 | Dose | 8.90 | Dose | 6.38 |
| 0.053 | 11.56 | 0.049 | 8.19 | 0.058 | 6.11 |
| 0.226 | 11.03 | 0.129 | 7.81 | 0.113 | 5.96 |
| 0.450 | 10.76 | 0.242 | 7.79 | 0.169 | 5.90 |
| 0.801 | 10.48 | 0.421 | 7.74 | 0.283 | 5.85 |
| 1.095 | 10.38 | 0.622 | 7.36 | 0.535 | 5.65 |
| 1.541 | 10.12 | 0.793 | 7.44 | 0.880 | 5.55 |
| 1.956 | 9.97 | 0.936 | 7.36 | 1.152 | 5.47 |
| 2.523 | 9.66 | 1.564 | 7.07 | 1.603 | 5.40 |
| 2.650 | 9.65 | 1.980 | 7.07 | 2.004 | 5.35 |
| 3.117 | 9.54 | 2.838 | 6.93 | 2.104 | 5.33 |
| 4.783 | 9.23 | 3.310 | 6.88 | 2.481 | 5.27 |
| 7.895 | 8.70 | 5.136 | 6.61 | 2.829 | 5.24 |
| 24.08 | 7.20 | 8.060 | 6.45 | 5.557 | 5.00 |
| 24.18 | 7.30 | 8.192 | 6.35 | 24.54 | 4.20 |
| 47.97 | 6.15 | 24.04 | 5.45 | 24.59 | 4.24 |
| 48.08 | 6.10 | 24.12 | 5.60 | 50.02 | 3.79 |
| 119.9 | 4.35 | 47.79 | 4.79 | 69.32 | 3.51 |
| 119.9 | 4.39 | 73.15 | 4.40 | 121.5 | 3.04 |
| | | 120.3 | 3.61 | 121.5 | 3.03 |
| | | | | 121.5 | 3.05 |

**Table A.12**
**NPNV EC temperature studies (2.5:1 $Cl_2$:TOC)**

| 4°C | | 30°C | |
|---|---|---|---|
| Time (hr) | $Cl_2$ residual (mg/L) | Time (hr) | $Cl_2$ residual (mg/L) |
| Dose | 8.56 | Dose | 8.56 |
| 0.046 | 7.55 | 0.053 | 7.31 |
| 0.123 | 7.45 | 0.122 | 7.10 |
| 0.197 | 7.37 | 0.165 | 7.00 |
| 0.281 | 7.30 | 0.253 | 6.95 |
| 0.524 | 7.15 | 0.503 | 6.60 |
| 0.736 | 7.07 | 0.767 | 6.56 |
| 1.488 | 6.96 | 0.994 | 6.48 |
| 2.048 | 6.90 | 1.516 | 6.22 |
| 2.523 | 6.90 | 2.153 | 6.10 |
| 3.148 | 6.83 | 2.280 | 6.06 |
| 5.023 | 6.63 | 2.517 | 6.00 |
| 5.173 | 6.71 | 2.991 | 5.91 |
| 8.553 | 6.51 | 3.094 | 5.89 |
| 8.667 | 6.58 | 5.268 | 5.49 |
| 24.89 | 6.19 | 7.752 | 5.18 |
| 25.00 | 6.25 | 23.98 | 3.65 |
| 49.18 | 5.75 | 24.15 | 3.68 |
| 75.28 | 5.71 | 47.94 | 2.62 |
| 122.6 | 5.35 | 74.22 | 2.22 |
| 122.6 | 5.42 | 121.6 | 1.53 |
| | | 121.6 | 1.54 |

**Table A.13**
**NPNV EC pH studies (1.5:1 $Cl_2$:TOC)**

| pH 5.5 | | pH 10.5 | | Changing pH (5.5 to 8.0) | |
|---|---|---|---|---|---|
| Time (hr) | $Cl_2$ residual (mg/L) | Time (hr) | $Cl_2$ residual (mg/L) | Time (hr) | $Cl_2$ residual (mg/L) |
| Dose | 5.39 | Dose | 5.39 | Dose | 5.35 |
| 0.044 | 4.48 | 0.037 | 4.77 | 0.041 | 4.97 |
| 0.097 | 4.38 | 0.091 | 4.65 | 0.195 | 4.80 |
| 0.168 | 4.24 | 0.153 | 4.58 | 0.326 | 4.70 |
| 0.234 | 4.16 | 0.257 | 4.50 | 0.512 | 4.64 |
| 0.466 | 4.04 | 0.544 | 4.17 | 0.745 | 4.57 |
| 0.729 | 3.95 | 0.779 | 4.21 | 1.969 | 4.35 |
| 1.000 | 3.80 | 1.168 | 3.92 | 3.252 | 4.05 |
| 1.546 | 3.75 | 1.262 | 3.91 | 3.362 | 4.11 |
| 2.031 | 3.65 | 1.706 | 3.76 | 4.896 | 3.96 |
| 2.193 | 3.64 | 2.081 | 3.75 | 4.959 | 4.04 |
| 2.567 | 3.60 | 2.594 | 3.65 | 5.207 | 3.79 |
| 3.005 | 3.55 | 2.659 | 3.62 | 5.301 | 3.75 |
| 5.334 | 3.31 | 3.026 | 3.55 | 7.889 | 3.21 |
| 8.766 | 3.19 | 4.842 | 3.25 | 7.971 | 3.25 |
| 23.40 | 2.68 | 8.475 | 2.93 | 25.07 | 2.40 |
| 23.44 | 2.76 | 22.52 | 2.05 | 25.14 | 2.42 |
| 71.51 | 2.06 | 22.59 | 2.03 | 47.37 | 1.98 |
| 119.7 | 1.75 | 71.11 | 0.93 | 72.26 | 1.58 |
| | | 71.34 | 0.89 | 119.6 | 0.99 |
| | | 71.34 | 0.89 | 119.6 | 0.99 |
| | | 119.3 | 0.42 | 119.6 | 1.05 |

**Table A.14**
**NPNV rechlorination studies**

| 24-hour | | 72-hour | |
|---|---|---|---|
| Time (hr) | $Cl_2$ residual (mg/L) | Time (hr) | $Cl_2$ residual (mg/L) |
| Dose | 2.26 | Dose | 2.73 |
| 0.063 | 1.75 | 0.057 | 2.08 |
| 0.288 | 1.52 | 0.530 | 1.79 |
| 0.544 | 1.37 | 1.046 | 1.60 |
| 1.269 | 1.18 | 3.053 | 1.35 |
| 3.274 | 0.95 | 13.94 | 0.81 |
| 3.333 | 0.93 | 23.97 | 0.55 |
| 7.958 | 0.77 | 24.01 | 0.56 |
| 24.38 | 0.32 | 50.18 | 0.28 |
| 24.47 | 0.40 | 67.07 | 0.21 |
| 50.52 | 0.10 | 67.13 | 0.10 |
| 73.13 | 0.00 | 91.54 | 0.01 |
| $ReCl_2$ | 3.04 | $ReCl_2$ | 2.60 |
| 24.80 | 3.40 | 67.50 | 2.76 |
| 1.578 | 3.30 | 67.54 | 2.68 |
| 2.078 | 3.25 | 68.02 | 2.59 |
| 2.560 | 3.25 | 68.55 | 2.57 |
| 27.42 | 2.50 | 91.62 | 2.06 |
| 50.06 | 2.17 | 91.69 | 2.06 |
| 69.71 | 1.80 | 109.1 | 1.60 |
| 95.34 | 1.55 | 136.2 | 1.45 |
| 95.39 | 1.45 | 159.9 | 1.25 |
| 120.5 | 1.40 | | |

**Table A.15**
**NPNV EC duplicate studies**

| 1.5:1 $Cl_2$:TOC dose | | 2.5:1 $Cl_2$:TOC dose | |
|---|---|---|---|
| Time (hr) | $Cl_2$ residual (mg/L) | Time (hr) | $Cl_2$ residual (mg/L) |
| Dose | 5.33 | Dose | 8.89 |
| 0.051 | 4.69 | 0.047 | 8.10 |
| 0.548 | 4.14 | 0.501 | 7.55 |
| 1.045 | 4.10 | 1.020 | 7.31 |
| 3.019 | 3.79 | 2.994 | 7.01 |
| 7.989 | 3.38 | 8.174 | 6.56 |
| 25.33 | 2.72 | 25.35 | 5.67 |
| 25.37 | 2.70 | 25.41 | 5.70 |
| 53.21 | 2.12 | 53.09 | 5.00 |
| 71.72 | 1.96 | 71.69 | 4.73 |
| 121.2 | 1.45 | 121.2 | 4.16 |
| | | 121.2 | 4.10 |

## Table A.16
## Winchester treatment studies

| EC (0.75:1 $Cl_2$:TOC) | | EC (1.5:1 $Cl_2$:TOC) | | UV (1.5:1 $Cl_2$:TOC) | | $O_3$ (1.5:1 $Cl_2$:TOC) | | $O_3$/BF (1.5:1 $Cl_2$:TOC) | |
|---|---|---|---|---|---|---|---|---|---|
| Time (hr) | $Cl_2$ residual (mg/L) | Time (hr) | $Cl_2$ residual (mg/L) | Time (hr) | $Cl_2$ residual (mg/L) | Time (hr) | $Cl_2$ residual (mg/L) | Time (hr) | $Cl_2$ residual (mg/L) |
| Dose | 1.58 | Dose | 3.08 | Dose | 2.65 | Dose | 2.58 | Dose | 2.17 |
| 0.069 | 1.30 | 0.040 | 2.72 | 0.043 | 2.25 | 0.046 | 2.28 | 0.057 | 2.00 |
| 0.125 | 1.22 | 0.145 | 2.65 | 0.120 | 2.14 | 0.113 | 2.21 | 0.138 | 2.00 |
| 0.198 | 1.17 | 0.229 | 2.57 | 0.191 | 2.10 | 0.172 | 2.14 | 0.176 | 1.98 |
| 0.280 | 1.13 | 0.272 | 2.54 | 0.275 | 2.07 | 0.247 | 2.13 | 0.264 | 1.93 |
| 0.553 | 1.02 | 0.547 | 2.42 | 0.543 | 2.01 | 0.516 | 2.01 | 0.547 | 1.86 |
| 0.786 | 0.94 | 0.799 | 2.36 | 0.812 | 1.90 | 0.750 | 1.94 | 0.778 | 1.80 |
| 1.020 | 0.93 | 0.977 | 2.34 | 1.073 | 1.88 | 1.124 | 1.92 | 1.001 | 1.78 |
| 1.076 | 0.92 | 1.031 | 2.32 | 1.141 | 1.87 | 1.168 | 1.90 | 1.093 | 1.77 |
| 1.798 | 0.87 | 1.459 | 2.29 | 1.698 | 1.79 | 1.627 | 1.80 | 1.429 | 1.77 |
| 2.566 | 0.86 | 2.072 | 2.23 | 1.940 | 1.78 | 2.100 | 1.80 | 2.283 | 1.70 |
| 2.984 | 0.80 | 2.539 | 2.20 | 2.642 | 1.71 | 2.514 | 1.76 | 2.514 | 1.70 |
| 3.042 | 0.79 | 3.014 | 2.18 | 3.093 | 1.70 | 3.017 | 1.71 | 2.937 | 1.69 |
| 5.069 | 0.70 | 3.079 | 2.19 | 3.207 | 1.69 | 3.074 | 1.70 | 2.998 | 1.68 |
| 22.22 | 0.45 | 5.063 | 2.06 | 4.998 | 1.59 | 4.967 | 1.63 | 5.519 | 1.60 |
| 22.26 | 0.40 | 18.47 | 1.73 | 13.34 | 1.28 | 12.22 | 1.30 | 8.999 | 1.44 |
| 38.67 | 0.30 | 26.28 | 1.50 | 24.38 | 1.15 | 23.57 | 1.22 | 23.92 | 1.25 |
| 65.75 | 0.15 | 26.45 | 1.55 | 24.49 | 1.15 | 23.61 | 1.21 | 23.01 | 1.25 |
| | | 37.28 | 1.30 | 47.18 | 0.86 | 45.91 | 0.93 | 45.71 | 1.00 |
| | | 64.87 | 1.15 | 71.54 | 0.67 | 70.22 | 0.68 | 76.12 | 0.80 |
| | | | | 122.0 | 0.44 | 120.4 | 0.51 | 122.8 | 0.64 |
| | | | | 122.0 | 0.42 | 120.4 | 0.55 | 122.8 | 0.59 |

**Table A.17**
**Winchester EC temperature studies (1.5:1 $Cl_2$:TOC)**

| 4°C | | 30°C | |
|---|---|---|---|
| Time (hr) | $Cl_2$ residual (mg/L) | Time (hr) | $Cl_2$ residual (mg/L) |
| Dose | 3.07 | Dose | 3.07 |
| 0.043 | 2.61 | 0.036 | 2.60 |
| 0.115 | 2.55 | 0.111 | 2.52 |
| 0.268 | 2.50 | 0.287 | 2.33 |
| 0.373 | 2.43 | 0.500 | 2.24 |
| 1.114 | 2.35 | 0.997 | 2.15 |
| 1.160 | 2.34 | 2.017 | 1.93 |
| 1.471 | 2.33 | 2.788 | 1.86 |
| 2.016 | 2.29 | 3.014 | 1.83 |
| 3.076 | 2.25 | 3.097 | 1.80 |
| 3.139 | 2.24 | 6.173 | 1.55 |
| 4.247 | 2.20 | 10.44 | 1.34 |
| 24.60 | 1.95 | 28.55 | 0.65 |
| 24.68 | 1.95 | 28.59 | 0.60 |
| 45.08 | 1.83 | 53.07 | 0.13 |
| 72.00 | 1.78 | 72.00 | 0.00 |

**Table A.18**
**Winchester EC pH studies (1.5:1 $Cl_2$:TOC)**

| pH 5.5 | | pH 10.5 | |
|---|---|---|---|
| Time (hr) | $Cl_2$ residual (mg/L) | Time (hr) | $Cl_2$ residual (mg/L) |
| Dose | 3.08 | Dose | 3.08 |
| 0.047 | 2.48 | 0.040 | 2.63 |
| 0.116 | 2.44 | 0.111 | 2.54 |
| 0.175 | 2.33 | 0.211 | 2.50 |
| 0.252 | 2.32 | 0.256 | 2.50 |
| 0.452 | 2.29 | 0.571 | 2.41 |
| 0.790 | 2.23 | 1.118 | 2.33 |
| 1.024 | 2.20 | 1.461 | 2.27 |
| 1.088 | 2.15 | 2.043 | 2.19 |
| 1.534 | 2.15 | 3.026 | 2.12 |
| 1.971 | 2.09 | 3.093 | 2.13 |
| 2.946 | 2.05 | 4.883 | 2.00 |
| 3.022 | 2.04 | 25.20 | 1.43 |
| 4.953 | 1.95 | 25.27 | 1.44 |
| 25.68 | 1.57 | 45.73 | 1.14 |
| 25.76 | 1.62 | 71.87 | 0.93 |
| 46.27 | 1.42 | | |
| 72.22 | 1.28 | | |

**Table A.19**
**Winchester rechlorination studies**

| 24-hour | | 48-hour | |
|---|---|---|---|
| Time (hr) | $Cl_2$ residual (mg/L) | Time (hr) | $Cl_2$ residual (mg/L) |
| Dose | 1.70 | Dose | 2.00 |
| 0.046 | 1.42 | 0.053 | 1.71 |
| 0.295 | 1.30 | 0.563 | 1.49 |
| 0.546 | 1.23 | 1.044 | 1.35 |
| 0.990 | 1.01 | 3.022 | 1.14 |
| 1.029 | 0.99 | 5.142 | 0.97 |
| 3.193 | 0.85 | 24.571 | 0.57 |
| 3.276 | 0.83 | 24.629 | 0.57 |
| 5.159 | 0.70 | 52.668 | 0.19 |
| 23.866 | 0.26 | 44.439 | 0.30 |
| 23.936 | 0.29 | 75.294 | 0.11 |
| 51.895 | 0.01 | $ReCl_2$ | 1.58 |
| $ReCl_2$ | 1.58 | 54.595 | 1.69 |
| 24.175 | 1.85 | 54.638 | 1.59 |
| 0.046 | 1.78 | 55.142 | 1.45 |
| 0.303 | 1.81 | 55.601 | 1.49 |
| 3.026 | 1.74 | 75.427 | 1.09 |
| 27.856 | 1.22 | 95.481 | 0.99 |
| 50.518 | 1.03 | 124.18 | 0.73 |
| 70.542 | 0.90 | 124.30 | 0.79 |
| 98.93 | 0.66 | 147.33 | 0.61 |
| 99.00 | 0.64 | 167.73 | 0.52 |
| 122.40 | 0.48 | 167.80 | 0.51 |
| 142.85 | 0.47 | | |
| 142.88 | 0.48 | | |

**Table A.20**
**Winchester EC duplicate study (1.5:1 $Cl_2$:TOC)**

| Time (hr) | $Cl_2$ residual (mg/L) | Time (hr) | $Cl_2$ residual (mg/L) |
|---|---|---|---|
| 0.000 | 3.40 | 52.98 | 1.52 |
| 0.049 | 3.04 | 75.65 | 1.45 |
| 1.130 | 2.82 | 124.3 | 1.02 |
| 3.039 | 2.58 | 147.7 | 0.87 |
| 24.90 | 2.00 | 168.1 | 0.85 |
| 24.97 | 1.99 | | |

**Table A.21**
**Sediment in tap water—chlorine decay**

| 0 g/L sediment | | 0.5 g/L sediment | | 1.0 g/L sediment | |
|---|---|---|---|---|---|
| Time (hr) | $Cl_2$ residual (mg/L) | Time (hr) | $Cl_2$ residual (mg/L) | Time (hr) | $Cl_2$ residual (mg/L) |
| Dose | 2.68 | Dose | 2.68 | Dose | 2.68 |
| 0.081 | 2.65 | 0.039 | 2.59 | 0.045 | 2.61 |
| 0.149 | 2.64 | 0.083 | Sediment | 0.092 | Sediment |
| 0.314 | 2.64 | 0.134 | 2.21 | 0.145 | 1.80 |
| 0.834 | 2.62 | 0.354 | 1.73 | 0.237 | 1.30 |
| 0.982 | 2.61 | 0.545 | 1.69 | 0.321 | 1.10 |
| 2.059 | 2.55 | 0.729 | 1.54 | 0.554 | 0.76 |
| 3.138 | 2.50 | 0.985 | 1.38 | 0.949 | 0.66 |
| 4.838 | 2.39 | 1.595 | 1.29 | 1.579 | 0.51 |
| 11.49 | 2.2 | 2.134 | 1.22 | 2.111 | 0.36 |
| 24.52 | 1.84 | 2.814 | 1.19 | 2.599 | 0.32 |
| 46.26 | 1.47 | 3.127 | 1.190 | 3.025 | 0.26 |
| 67.92 | 1.25 | 4.867 | 1.12 | 4.566 | 0.23 |
| 118.5 | 0.85 | 11.00 | 0.88 | 9.927 | 0.10 |
| | | 24.06 | 0.655 | 23.87 | 0.00 |
| | | 45.84 | 0.26 | | |
| | | 67.55 | 0.350 | | |
| | | 118.0 | 0.08 | | |

# APPENDIX B
# CHLORAMINE DECAY RAW DATA

**Table B.1**
**Measured experimental conditions table**

| | | |
|---|---|---|
| pH | Base case | 8.0 ± 0.1 |
| | High pH | 10.5 ± 0.2 |
| Temperature | Base case | 20°C ± 2°C |
| | Low temp | 4°C ± 2°C |
| | High temp | 30°C ± 2°C |

Only non-base case conditions noted in tables

**Table B.2**
**Betasso treatment studies (2 mg/L CLM residual at 24 hr)**

| Raw | | EC | | GAC | | UV | | $O_3$/BF | |
|---|---|---|---|---|---|---|---|---|---|
| Time (hr) | CLM residual (mg/L) | Time (hr) | CLM residual (mg/L) | Time (hr) | CLM residual (mg/L) | Time (hr) | CLM residual (mg/L) | Time (hr) | CLM residual (mg/L) |
| Dose | 2.91 | Dose | 2.46 | Dose | 2.20 | Dose | 2.49 | Dose | 2.25 |
| 0.27 | 2.63 | 0.24 | 2.41 | 0.34 | 2.00 | 0.25 | 2.10 | 0.28 | 2.24 |
| 0.50 | 2.62 | 0.52 | 2.20 | 0.52 | 1.90 | 0.51 | 2.27 | 0.56 | 2.25 |
| 0.95 | 2.66 | 0.99 | 2.36 | 1.07 | 1.97 | 0.98 | 2.29 | 0.99 | 2.27 |
| 1.99 | 2.60 | 1.06 | 2.37 | 2.01 | 1.95 | 1.03 | 2.25 | 1.99 | 2.25 |
| 4.72 | 2.58 | 1.91 | 2.34 | 4.16 | 1.91 | 1.97 | 2.25 | 4.03 | 2.23 |
| 8.03 | 2.50 | 4.02 | 2.30 | 8.12 | 1.90 | 3.97 | 2.21 | 7.99 | 2.21 |
| 24.00 | 2.40 | 8.07 | 2.27 | 25.39 | 1.96 | 8.05 | 2.20 | 24.01 | 2.17 |
| 48.80 | 2.25 | 24.21 | 2.18 | 46.78 | 1.93 | 24.40 | 1.97 | 51.28 | 2.15 |
| 72.11 | 2.23 | 47.98 | 1.87 | 79.03 | 1.93 | 24.55 | 2.00 | 74.09 | 2.06 |
| 120.1 | 2.12 | 72.79 | 2.15 | 257.3 | 1.70 | 48.12 | 1.95 | 126.3 | 2.10 |
| 240.9 | 1.90 | 128.0 | 2.05 | 528.6 | 1.63 | 72.35 | 1.95 | 269.6 | 1.84 |
| 485.6 | 1.50 | 239.7 | 1.81 | | | 72.47 | 1.95 | 484.6 | 1.40 |
| | | 477.0 | 1.73 | | | 120.1 | 2.00 | | |
| | | | | | | 120.1 | 2.00 | | |
| | | | | | | 240.3 | 1.84 | | |
| | | | | | | 481.6 | 1.57 | | |

**Table B.3**
**Betasso treatment studies (3 mg/L CLM residual at 24 hr)**

| Raw | | EC | | GAC | | $O_3$/BF | |
|---|---|---|---|---|---|---|---|
| Time (hr) | CLM residual (mg/L) | Time (hr) | CLM residual (mg/L) | Time (hr) | CLM residual (mg/L) | Time (hr) | CLM residual (mg/L) |
| Dose | 4.28 | Dose | 3.54 | Dose | 3.29 | Dose | 3.25 |
| 0.26 | 3.65 | 0.27 | 3.25 | 0.25 | 3.05 | 0.26 | 3.20 |
| 0.52 | 3.65 | 0.51 | 3.30 | 0.51 | 3.10 | 0.52 | 3.21 |
| 1.06 | 3.65 | 1.01 | 3.30 | 1.01 | 3.11 | 0.97 | 3.27 |
| 1.94 | 3.61 | 1.97 | 3.20 | 1.95 | 3.10 | 1.94 | 3.21 |
| 4.17 | 3.63 | 2.03 | 3.20 | 4.01 | 3.05 | 4.00 | 3.20 |
| 8.05 | 3.55 | 3.99 | 3.20 | 7.97 | 3.09 | 7.95 | 3.20 |
| 23.79 | 3.46 | 8.02 | 3.23 | 25.41 | 3.15 | 24.09 | 3.16 |
| 48.24 | 3.32 | 23.97 | 3.16 | 46.73 | 3.00 | 51.21 | 3.15 |
| 71.96 | 3.26 | 47.99 | 3.07 | 78.99 | 3.06 | 73.93 | 3.02 |
| 119.6 | 3.03 | 71.98 | 2.80 | 121.4 | 3.06 | 126.3 | 2.94 |
| 240.3 | 2.83 | 131.2 | 2.61 | 257.2 | 2.90 | 269.4 | 2.74 |
| 484.0 | 2.31 | 240.0 | 2.83 | 528.5 | 2.75 | 484.6 | 2.57 |
| | | 480.0 | 2.13 | | | | |

**Table B.4**
**Betasso treatment studies (4 mg/L CLM residual at 24 hr)**

| Raw | | EC | | GAC | | UV | | $O_3$/BF | |
|---|---|---|---|---|---|---|---|---|---|
| Time (hr) | CLM residual (mg/L) | Time (hr) | CLM residual (mg/L) | Time (hr) | CLM residual (mg/L) | Time (hr) | CLM residual (mg/L) | Time (hr) | CLM residual (mg/L) |
| Dose | 5.77 | Dose | 5.60 | Dose | 4.50 | Dose | 5.60 | Dose | 4.50 |
| 0.28 | 4.82 | 0.25 | 4.84 | 0.25 | 4.35 | 0.260 | 4.80 | 0.28 | 4.26 |
| 0.49 | 4.91 | 0.49 | 4.99 | 0.66 | 4.25 | 0.491 | 4.73 | 0.50 | 4.34 |
| 1.06 | 4.86 | 1.08 | 4.66 | 1.01 | 4.29 | 1.020 | 4.78 | 1.04 | 4.33 |
| 1.98 | 4.90 | 1.80 | 5.09 | 2.11 | 4.30 | 1.963 | 4.65 | 2.12 | 4.38 |
| 4.09 | 4.80 | 4.04 | 4.95 | 3.96 | 4.35 | 3.796 | 4.65 | 4.00 | 4.34 |
| 7.99 | 4.79 | 7.96 | 5.00 | 7.84 | 4.25 | 7.829 | 4.68 | 7.95 | 4.32 |
| 23.87 | 4.66 | 24.14 | 4.85 | 25.21 | 4.36 | 24.09 | 4.64 | 24.15 | 4.27 |
| 48.13 | 4.38 | 47.91 | 4.70 | 46.38 | 4.34 | 47.81 | 4.51 | 51.21 | 4.18 |
| 71.98 | 4.43 | 72.77 | 4.75 | 78.68 | 4.28 | 72.26 | 4.55 | 73.89 | 4.05 |
| 119.6 | 4.25 | 127.9 | 4.65 | 121.0 | 4.22 | 120.0 | 4.44 | 126.3 | 4.00 |
| 240.2 | 3.98 | 238.6 | 4.48 | 256.8 | 4.05 | 240.0 | 4.22 | 269.4 | 3.75 |
| 485 | 3.17 | 477 | 4.10 | 528.1 | 3.80 | 480 | 3.85 | 484.5 | 3.50 |

**Table B.5**
**Betasso temperature studies (3 mg/L CLM residual at 24 hr)**

| EC 4°C | | EC 30°C | | $O_3$/BF 4°C | | $O_3$/BF 30°C | |
|---|---|---|---|---|---|---|---|
| Time (hr) | CLM residual (mg/L) | Time (hr) | CLM residual (mg/L) | Time (hr) | CLM residual (mg/L) | Time (hr) | CLM residual (mg/L) |
| Dose | 3.54 | Dose | 3.54 | Dose | 3.22 | Dose | 3.23 |
| 0.27 | 3.33 | 0.26 | 3.01 | 0.27 | 2.78 | 0.25 | 2.70 |
| 0.50 | 3.18 | 0.51 | 3.08 | 0.59 | 2.76 | 0.50 | 2.67 |
| 1.02 | 3.25 | 0.98 | 3.08 | 0.96 | 2.77 | 1.14 | 2.73 |
| 1.99 | 3.23 | 2.04 | 3.05 | 1.94 | 2.75 | 2.09 | 2.55 |
| 4.13 | 3.22 | 4.12 | 3.01 | 3.83 | 2.75 | 3.90 | 2.65 |
| 7.56 | 3.21 | 8.01 | 2.80 | 7.85 | 2.75 | 8.13 | 2.60 |
| 24.06 | 3.16 | 23.87 | 2.82 | 24.51 | 2.75 | 24.51 | 2.50 |
| 48.01 | 3.05 | 48.03 | 2.79 | 48.28 | 2.75 | 50.45 | 2.29 |
| 70.45 | 3.13 | 72.12 | 2.64 | 70.68 | 2.70 | 72.88 | 2.26 |
| 120.38 | 3.01 | 145.53 | 2.41 | 128.00 | 2.73 | 125.33 | 2.08 |
| 480.00 | 2.70 | 336.00 | 2.25 | 266.12 | 2.60 | 268.14 | 1.79 |
| | | 480.00 | 1.91 | 481.27 | 2.13 | 483.48 | 1.57 |

**Table B.6**
**Betasso pH studies (3 mg/L CLM residual at 24 hr)**

| GAC pH 10.5 | | EC pH 10.5 | |
|---|---|---|---|
| Time (hr) | CLM residual (mg/L) | Time (hr) | CLM residual (mg/L) |
| Dose | 3.20 | Dose | 3.04 |
| 0.31 | 3.16 | 0.24 | 2.97 |
| 0.46 | 3.20 | 0.48 | 3.02 |
| 0.92 | 3.09 | 0.96 | 2.97 |
| 2.00 | 3.15 | 1.94 | 2.85 |
| 3.87 | 3.10 | 3.96 | 2.77 |
| 6.29 | 3.09 | 7.97 | 2.62 |
| 25.61 | 2.54 | 24.73 | 2.27 |
| 48.30 | 2.33 | 49.88 | 2.25 |
| 71.50 | 2.45 | 73.14 | 2.33 |
| 99.59 | 2.31 | 124.82 | 2.20 |
| 239.92 | 2.25 | 240.25 | 2.10 |
| 552.42 | 1.95 | 481.25 | 1.75 |

**Table B.7**
**Betasso blending studies**

| 63rd St EC different residual | | 63rd St EC different demand | | Betasso EC 2 hour / 5 day | | Betasso EC 2 hour / 24 hour | |
|---|---|---|---|---|---|---|---|
| Time (hr) | CLM residual (mg/L) | Time (hr) | CLM residual (mg/L) | Time (hr) | CLM residual (mg/L) | Time (hr) | CLM residual (mg/L) |
| Dose | 5.57 | Dose | 5.6 | Dose | 3.29 | Dose | 3.29 |
| 0.22 | 3.35 | 0.2 | 3.8 | 1.04 | 2.87 | 1.22 | 2.84 |
| 1.70 | 3.31 | 1.8 | 3.8 | 2.69 | 2.85 | 2.38 | 2.85 |
| | | 24.0 | 3.7 | 24.00 | 2.80 | 23.63 | 2.82 |
| | | | | 51.33 | 2.77 | | |
| | | | | 71.76 | 2.76 | | |
| | | | | 118.9 | 2.70 | | |

| Betasso GAC | | Betasso GAC | | Betasso GAC 2 hr | | Betasso GAC 2 hr | |
|---|---|---|---|---|---|---|---|
| Time (hr) | CLM residual (mg/L) | Time (hr) | CLM residual (mg/L) | Time (hr) | CLM residual (mg/L) | Time (hr) | CLM residual (mg/L) |
| Dose | 2.21 | Dose | 4.2 | 119.88 | 3.29 | 23.27 | 3.29 |
| 0.23 | 1.93 | 0.2 | 3.5 | 121.07 | 2.75 | 24.39 | 2.95 |
| 1.75 | 1.90 | 1.8 | 3.5 | 122.11 | 2.78 | 25.32 | 3.05 |
| | | 24.0 | 3.3 | | | | |

| Blended water | | Blended water | | Blended water | | Blended water | |
|---|---|---|---|---|---|---|---|
| Time (hr) | CLM residual (mg/L) | Time (hr) | CLM residual (mg/L) | Time (hr) | CLM residual (mg/L) | Time (hr) | CLM residual (mg/L) |
| 1.89 | 2.55 | 1.93 | 3.63 | 4.0 | 2.65 | 30.825 | 3 |
| 3.82 | 2.56 | 4.17 | 3.40 | 7.9 | 2.70 | 48.62 | 2.86 |
| 6.99 | 2.53 | 8.34 | 3.51 | 7.9 | 2.70 | 71.24 | 2.84 |
| 23.20 | 2.48 | 8.38 | 3.54 | 26.1 | 2.63 | 97.67 | 2.73 |
| 23.31 | 2.50 | 24.17 | 3.42 | 26.1 | 2.63 | 145.1 | 2.63 |
| 47.27 | 2.40 | 24.27 | 3.41 | 48.0 | 2.58 | 266.8 | 2.50 |
| 70.10 | 2.38 | 48.71 | 3.33 | 70.6 | 2.54 | 791.6 | 2.10 |
| 116.8 | 2.25 | 75.46 | 3.26 | 114.8 | 2.51 | | |
| 116.8 | 2.25 | 120.4 | 2.95 | 114.9 | 2.52 | | |
| | | 120.5 | 3.07 | 479.7 | 1.95 | | |

**Table B.8**
**Betasso EC duplicate study (3 mg/L CLM residual at 24 hr)**

| Time (hr) | CLM residual (mg/L) | Time (hr) | CLM residual (mg/L) |
|---|---|---|---|
| Dose | 3.58 | 24.07 | 3.20 |
| 0.26 | 3.40 | 48.12 | 3.20 |
| 0.47 | 3.44 | 76.57 | 3.20 |
| 0.98 | 3.45 | 124.6 | 3.05 |
| 2.45 | 3.40 | 294.5 | 2.85 |
| 4.24 | 3.36 | 509.7 | 2.54 |
| 8.00 | 3.34 | | |

**Table B.9**
**NPNV treatment studies (2 mg/L CLM residual at 24 hr)**

| Raw | | EC | | GAC | | UV | | $O_3$/BF | |
|---|---|---|---|---|---|---|---|---|---|
| Time (hr) | CLM residual (mg/L) | Time (hr) | CLM residual (mg/L) | Time (hr) | CLM residual (mg/L) | Time (hr) | CLM residual (mg/L) | Time (hr) | CLM residual (mg/L) |
| Dose | 3.24 | Dose | 3.01 | Dose | 2.89 | Dose | 3.01 | Dose | 2.86 |
| 0.20 | 2.67 | 0.29 | 3.04 | 0.26 | 2.73 | 0.19 | 2.95 | 0.50 | 2.85 |
| 0.44 | 2.65 | 0.65 | 2.95 | 0.51 | 2.71 | 0.48 | 2.87 | 1.04 | 2.87 |
| 0.95 | 2.55 | 1.01 | 2.95 | 0.99 | 2.65 | 0.97 | 2.88 | 1.98 | 2.85 |
| 0.98 | 2.58 | 1.89 | 2.95 | 1.06 | 2.65 | 1.02 | 2.85 | 4.61 | 2.80 |
| 2.11 | 2.55 | 1.92 | 2.94 | 1.98 | 2.66 | 2.01 | 2.93 | 15.75 | 2.74 |
| 4.29 | 2.50 | 3.27 | 2.95 | 4.06 | 2.67 | 3.86 | 2.85 | 27.56 | 2.50 |
| 7.86 | 2.35 | 8.05 | 2.8 | 8.82 | 2.60 | 24.12 | 2.75 | 47.39 | 2.58 |
| 24.00 | 2.28 | 23.79 | 2.87 | 23.94 | 2.38 | 24.18 | 2.70 | 47.44 | 2.58 |
| 24.10 | 2.25 | 23.84 | 2.78 | 23.92 | 2.39 | 46.96 | 2.61 | 72.63 | 2.50 |
| 49.15 | 2.06 | 50.54 | 2.65 | 48.77 | 2.30 | 98.37 | 2.70 | 72.68 | 2.55 |
| 71.91 | 1.89 | 72.40 | 2.47 | 71.50 | 2.29 | 98.44 | 2.67 | 125.56 | 2.19 |
| 72.03 | 1.86 | 72.54 | 2.57 | 71.55 | 2.27 | 118.43 | 2.58 | 125.44 | 2.16 |
| 121.4 | 1.71 | 121.4 | 2.38 | 122.5 | 2.15 | 240.0 | 2.27 | 220.5 | 1.98 |
| 121.4 | 1.68 | 121.6 | 2.32 | 122.6 | 2.13 | 503.7 | 1.87 | 499.0 | 1.58 |
| 240.4 | 1.35 | 237.6 | 2.05 | 241.6 | 1.72 | 503.6 | 1.93 | 498.9 | 1.60 |
| 603.5 | 0.80 | 480.9 | 1.65 | 482.1 | 1.35 | | | | |

**Table B.10**
**NPNV treatment studies (4 mg/L CLM residual at 24 hr)**

| Raw | | EC | | GAC | | UV | | $O_3$/BF | |
|---|---|---|---|---|---|---|---|---|---|
| Time (hr) | CLM residual (mg/L) | Time (hr) | CLM residual (mg/L) | Time (hr) | CLM residual (mg/L) | Time (hr) | CLM residual (mg/L) | Time (hr) | CLM residual (mg/L) |
| Dose | 6.45 | Dose | 5.29 | Dose | 5.02 | Dose | 5.30 | Dose | 5.02 |
| 0.22 | 5.35 | 0.26 | 4.71 | 0.30 | 4.59 | 0.20 | 5.20 | 0.26 | 4.45 |
| 0.44 | 5.32 | 0.65 | 4.69 | 0.51 | 4.60 | 0.46 | 5.15 | 0.55 | 4.43 |
| 0.99 | 5.36 | 1.00 | 4.60 | 1.10 | 4.58 | 0.96 | 5.10 | 1.01 | 4.41 |
| 2.11 | 5.35 | 1.89 | 4.65 | 1.14 | 4.60 | 1.00 | 5.05 | 1.86 | 4.40 |
| 4.29 | 5.23 | 1.93 | 4.60 | 1.96 | 4.53 | 2.03 | 5.05 | 4.58 | 4.35 |
| 7.85 | 5.15 | 3.23 | 4.55 | 4.07 | 4.55 | 3.83 | 5.00 | 15.58 | 4.24 |
| 24.17 | 4.90 | 8.09 | 4.45 | 8.79 | 4.35 | 24.05 | 4.95 | 27.53 | 4.20 |
| 24.28 | 4.90 | 23.76 | 4.60 | 23.97 | 4.22 | 24.09 | 4.90 | 72.64 | 4.04 |
| 49.32 | 4.60 | 50.59 | 4.18 | 24.01 | 4.25 | 47.17 | 4.74 | 72.67 | 4.05 |
| 72.14 | 4.35 | 72.77 | 4.04 | 48.76 | 4.01 | 98.29 | 4.93 | 125.4 | 3.56 |
| 72.28 | 4.40 | 72.83 | 4.05 | 71.54 | 3.95 | 98.34 | 4.90 | 125.3 | 3.55 |
| 120.8 | 4.10 | 121.4 | 3.75 | 71.59 | 4.00 | 118.6 | 4.80 | 220.6 | 3.24 |
| 120.8 | 4.10 | 238.2 | 3.32 | 122.4 | 3.75 | 240.2 | 4.28 | 500.4 | 2.80 |
| 240.4 | 3.45 | 480.9 | 2.78 | 122.4 | 3.75 | 503.7 | 3.75 | 500.3 | 2.85 |
| 603.5 | 2.39 | | | 241.5 | 3.07 | 503.7 | 3.79 | | |
| | | | | 482.2 | 2.60 | | | | |

**Table B.11**
**NPNV $O_3$ treatment studies**

| 2 mg/L CLM residual at 24 hr | | 4 mg/L CLM residual at 24 hr | |
|---|---|---|---|
| Time (hr) | CLM residual (mg/L) | Time (hr) | CLM residual (mg/L) |
| Dose | 2.99 | Dose | 5.28 |
| 0.26 | 2.95 | 0.22 | 5.00 |
| 0.48 | 2.68 | 0.44 | 4.75 |
| 0.95 | 2.65 | 0.89 | 4.73 |
| 3.95 | 2.65 | 1.90 | 4.63 |
| 3.74 | 2.55 | 3.68 | 4.55 |
| 12.65 | 2.32 | 12.59 | 4.35 |
| 23.64 | 2.15 | 23.60 | 4.32 |
| 23.67 | 2.10 | 23.64 | 4.18 |
| 47.80 | 1.79 | 47.77 | 3.81 |
| 71.29 | 1.56 | 71.33 | 3.45 |
| 71.36 | 1.55 | 71.38 | 3.46 |
| 120.17 | 1.31 | 119.92 | 3.10 |
| 231.48 | 1.07 | 231.50 | 2.77 |
| 501.97 | 0.80 | 476.08 | 2.20 |
| 501.08 | 0.85 | 476.17 | 2.18 |

**Table B.12**
**NPNV temperature studies (2 mg/L CLM residual at 24 hr)**

| EC 4°C | | EC 30°C | | $O_3$/BF 4°C | | $O_3$/BF 30°C | |
|---|---|---|---|---|---|---|---|
| Time (hr) | CLM residual (mg/L) | Time (hr) | CLM residual (mg/L) | Time (hr) | CLM residual (mg/L) | Time (hr) | CLM residual (mg/L) |
| Dose | 2.99 | Dose | 2.99 | Dose | 2.86 | Dose | 2.86 |
| 0.28 | 2.65 | 0.279 | 2.74 | 0.20 | 2.86 | 0.21 | 2.81 |
| 0.53 | 2.65 | 0.511 | 2.66 | 0.47 | 2.85 | 0.50 | 2.81 |
| 1.01 | 2.65 | 0.999 | 2.65 | 0.92 | 2.86 | 0.98 | 2.80 |
| 1.14 | 2.65 | 1.124 | 2.66 | 1.86 | 2.87 | 1.86 | 2.80 |
| 2.12 | 2.70 | 2.079 | 2.63 | 3.76 | 2.84 | 3.78 | 2.75 |
| 4.05 | 2.65 | 3.996 | 2.55 | 14.66 | 2.81 | 14.63 | 2.70 |
| 8.10 | 2.61 | 8.131 | 2.48 | 26.56 | 2.80 | 26.45 | 2.43 |
| 24.43 | 2.60 | 24.261 | 2.35 | 46.92 | 2.78 | 46.60 | 2.28 |
| 24.43 | 2.55 | 24.342 | 2.30 | 46.95 | 2.75 | 46.68 | 2.30 |
| 48.02 | 2.53 | 47.978 | 2.08 | 71.30 | 2.75 | 71.32 | 2.14 |
| 77.43 | 2.52 | 77.290 | 1.84 | 71.33 | 2.75 | 71.38 | 2.15 |
| 77.53 | 2.52 | 77.384 | 1.85 | 114.6 | 2.49 | 114.4 | 1.78 |
| 127.0 | 2.40 | 120.1 | 1.7 | 114.5 | 2.46 | 114.4 | 1.75 |
| 127.0 | 2.43 | 242.4 | 1.3 | 219.4 | 2.43 | 219.4 | 1.46 |
| 266.2 | 2.35 | 483.0 | 0.8 | 499.6 | 2.25 | 475.0 | 1.08 |
| 482.9 | 2.12 | | | 499.7 | 2.25 | 475.4 | 1.02 |

**Table B.13**
**NPNV EC pH studies (2 mg/L CLM residual at 24 hr)**

| pH 10.5 | |
|---|---|
| Time (hr) | CLM residual (mg/L) |
| Dose | 2.99 |
| 0.40 | 2.80 |
| 0.55 | 2.71 |
| 1.13 | 2.75 |
| 1.22 | 2.75 |
| 2.19 | 2.70 |
| 4.11 | 2.66 |
| 8.08 | 2.59 |
| 23.88 | 2.55 |
| 24.05 | 2.55 |
| 48.08 | 2.31 |
| 77.64 | 2.22 |
| 77.73 | 2.23 |
| 120.05 | 2.15 |
| 120.08 | 2.05 |
| 241.83 | 1.87 |
| 482.83 | 1.50 |

**Table B.14**
**NPNV different residual blending study**

| 63rd St EC | | EC NPNV | | Blended water | |
|---|---|---|---|---|---|
| Time (hr) | CLM residual (mg/L) | Time (hr) | CLM residual (mg/L) | Time (hr) | CLM residual (mg/L) |
| Dose | 2.36 | Dose | 4.96 | 2.21 | 3.82 |
| 0.29 | 2.28 | 0.24 | 4.50 | 3.96 | 3.22 |
| 0.48 | 2.30 | 0.46 | 4.45 | 16.88 | 3.10 |
| 1.01 | 2.18 | 0.98 | 4.48 | 24.09 | 2.98 |
| 1.86 | 2.22 | 1.78 | 4.42 | 24.14 | 3.10 |
| 2.05 | 2.24 | 2.13 | 4.35 | 48.40 | 3.20 |
| 3.88 | 2.15 | 3.88 | 4.30 | 71.21 | 3.00 |
| 16.81 | 2.15 | 16.92 | 4.10 | 124.58 | 2.95 |
| 23.98 | 2.10 | 23.95 | 4.20 | 124.63 | 2.91 |
| 24.02 | 2.07 | 23.98 | 4.23 | 238.85 | 2.85 |
| 48.32 | 2.04 | 48.36 | 4.03 | 475.17 | 2.45 |
| 71.14 | 2.06 | 71.14 | 3.78 | 475.22 | 2.45 |
| 119.92 | 2.03 | 120.00 | 3.78 | | |
| 119.97 | 2.00 | 120.03 | 3.78 | | |
| 234.30 | 2.00 | 234.18 | 3.70 | | |
| 477.00 | 1.80 | 477.67 | 3.05 | | |
| 477.03 | 1.76 | 477.63 | 3.11 | | |

**Table B.15**
**NPNV different demand blending study**

| 63rd St EC | | EC NPNV | | Blended water | |
|---|---|---|---|---|---|
| Time (hr) | CLM residual (mg/L) | Time (hr) | CLM residual (mg/L) | Time (hr) | CLM residual (mg/L) |
| Dose | 4.8 | Dose | 4.96 | 2.65 | 4.20 |
| 0.20 | 4.3 | 0.18 | 4.40 | 3.67 | 4.18 |
| 0.50 | 4.3 | 0.52 | 4.48 | 6.55 | 4.20 |
| 1.01 | 4.2 | 1.02 | 4.50 | 24.12 | 4.20 |
| 2.31 | 4.3 | 2.32 | 4.45 | 24.15 | 4.20 |
| 3.67 | 4.3 | 2.75 | 4.45 | 48.08 | 4.20 |
| 6.50 | 4.3 | 3.70 | 4.40 | 75.16 | 3.98 |
| 24.19 | 4.1 | 6.56 | 4.30 | 121.08 | 3.95 |
| 24.26 | 4.2 | 24.32 | 4.25 | 121.13 | 3.95 |
| 48.02 | 4.1 | 48.12 | 4.10 | 248.47 | 3.64 |
| 75.20 | 4.1 | 75.23 | 3.90 | 478.47 | 3.56 |
| 120.97 | 3.9 | 121.17 | 3.75 | 478.52 | 3.57 |
| 121.00 | 4.0 | 121.20 | 3.80 | | |
| 248.52 | 4.1 | 248.60 | 3.55 | | |
| 478.37 | 3.8 | 478.50 | 3.24 | | |
| 478.40 | 3.8 | 478.53 | 3.21 | | |

**Table B.16**
**NPNV rechloramination study**

| EC BC | | Post-addition CLM | | HD CLM solution | |
|---|---|---|---|---|---|
| Time (hr) | CLM residual (mg/L) | Time (hr) | CLM residual (mg/L) | Time (hr) | CLM residual (mg/L) |
| Dose | 3.02 | 171.08 | 3.07 | Dose | 4.02 |
| 0.27 | 2.74 | 171.15 | 3.07 | 0.07 | 3.98 |
| 0.44 | 2.80 | 171.22 | 3.04 | 0.14 | 3.90 |
| 0.96 | 2.78 | 193.10 | 2.99 | 0.93 | 3.80 |
| 2.46 | 2.78 | 236.02 | 2.90 | 15.92 | 3.60 |
| 7.24 | 2.70 | 236.00 | 2.82 | 23.42 | 3.45 |
| 26.66 | 2.40 | 288.25 | 2.75 | 23.53 | 3.47 |
| 26.74 | 2.50 | 288.28 | 2.75 | 168.13 | 2.90 |
| 96.21 | 2.26 | 410.50 | 2.55 | 236.67 | 2.78 |
| 121.97 | 2.18 | 411.00 | 2.55 | 237.30 | 2.80 |
| 169.20 | 2.07 | 649.58 | 2.20 | 407.77 | 2.40 |
| 169.25 | 2.05 | 649.67 | 2.24 | 470.92 | 2.10 |
| 236.88 | 2.00 | | | 470.98 | 2.20 |
| 411.08 | 1.83 | | | 670.22 | 2.05 |
| 480.83 | 1.69 | | | 670.30 | 2.10 |
| 480.87 | 1.67 | | | | |
| 649.25 | 1.45 | | | | |
| 649.37 | 1.50 | | | | |

**Table B.17**
**NPNV EC duplicate study (2 mg/L CLM residual at 24 hr)**

| Time (hr) | CLM residual (mg/L) |
|---|---|
| Dose | 2.99 |
| 0.24 | 2.85 |
| 0.48 | 2.80 |
| 1.00 | 2.78 |
| 1.07 | 2.80 |
| 1.99 | 2.81 |
| 4.02 | 2.84 |
| 7.33 | 2.60 |
| 25.09 | 2.59 |
| 25.13 | 2.54 |
| 52.00 | 2.48 |
| 119.9 | 2.35 |
| 119.9 | 2.30 |
| 240.18 | 2.05 |
| 506.88 | 1.45 |

**Table B.18**
**Winchester treatment studies (2 mg/L CLM residual at 24 hr)**

| EC | | GAC | | UV | | $O_3$/BF | | $O_3$ | |
|---|---|---|---|---|---|---|---|---|---|
| Time (hr) | CLM residual (mg/L) | Time (hr) | CLM residual (mg/L) | Time (hr) | CLM residual (mg/L) | Time (hr) | CLM residual (mg/L) | Time (hr) | CLM residual (mg/L) |
| Dose | 2.52 | Dose | 2.49 | Dose | 2.50 | Dose | 2.50 | Dose | 2.50 |
| 0.30 | 2.30 | 0.21 | 2.28 | 0.26 | 2.25 | 0.23 | 2.25 | 0.23 | 2.30 |
| 0.65 | 2.35 | 0.58 | 2.20 | 0.66 | 2.24 | 0.48 | 2.25 | 0.49 | 2.32 |
| 1.12 | 2.35 | 1.43 | 2.25 | 1.03 | 2.25 | 1.00 | 2.30 | 1.01 | 2.30 |
| 2.00 | 2.35 | 2.46 | 2.25 | 1.50 | 2.20 | 2.94 | 2.25 | 2.04 | 2.25 |
| 3.82 | 2.36 | 5.43 | 2.23 | 17.95 | 2.16 | 3.94 | 2.20 | 3.58 | 2.31 |
| 17.17 | 2.26 | 24.97 | 2.18 | 26.03 | 2.10 | 8.94 | 2.22 | 8.20 | 2.20 |
| 23.46 | 2.26 | 25.02 | 2.16 | 26.10 | 2.10 | 23.72 | 2.22 | 25.74 | 2.03 |
| 23.52 | 2.26 | 50.37 | 2.13 | 48.65 | 2.05 | 23.74 | 2.18 | 25.77 | 2.05 |
| 49.11 | 2.21 | 69.50 | 2.10 | 71.66 | 2.04 | 48.28 | 2.15 | 48.55 | 1.97 |
| 71.67 | 2.17 | 69.68 | 2.10 | 71.71 | 2.03 | 71.68 | 2.10 | 72.49 | 1.90 |
| 71.71 | 2.13 | 117.50 | 2.03 | 120.13 | 1.96 | 71.72 | 2.09 | 119.10 | 1.80 |
| 119.13 | 2.08 | 221.12 | 1.90 | 240.63 | 1.97 | 118.33 | 1.95 | 238.92 | 1.68 |
| 240.40 | 1.97 | 221.17 | 1.94 | 475.63 | 1.63 | 239.40 | 1.87 | 238.93 | 1.84 |
| 240.50 | 1.93 | 477.50 | 2.05 | 475.70 | 1.64 | 239.42 | 1.90 | 473.20 | 1.38 |
| 477.63 | 1.71 | 477.53 | 2.05 | | | 507.72 | 1.69 | 473.17 | 1.39 |
| 477.70 | 1.75 | | | | | 507.75 | 1.71 | | |

**Table B.19**
**Winchester treatment studies (4 mg/L CLM residual at 24 hr)**

| EC | | GAC | | UV | | $O_3$/BF | | $O_3$ | |
|---|---|---|---|---|---|---|---|---|---|
| Time (hr) | CLM residual (mg/L) | Time (hr) | CLM residual (mg/L) | Time (hr) | CLM residual (mg/L) | Time (hr) | CLM residual (mg/L) | Time (hr) | CLM residual (mg/L) |
| Dose | 4.81 | Dose | 4.79 | Dose | 4.84 | Dose | 4.80 | Dose | 4.81 |
| 0.28 | 4.26 | 0.24 | 4.35 | 0.27 | 4.37 | 0.24 | 4.30 | 0.21 | 4.40 |
| 0.52 | 4.20 | 0.55 | 4.18 | 0.66 | 4.00 | 0.47 | 4.23 | 0.45 | 4.31 |
| 1.01 | 4.26 | 1.45 | 4.38 | 1.00 | 4.02 | 0.95 | 4.27 | 0.95 | 4.28 |
| 1.97 | 4.20 | 2.71 | 4.30 | 1.48 | 4.10 | 1.97 | 4.20 | 1.99 | 4.30 |
| 3.69 | 4.22 | 5.54 | 4.32 | 17.94 | 3.88 | 3.89 | 4.24 | 3.54 | 4.22 |
| 17.05 | 4.19 | 25.15 | 4.28 | 26.07 | 3.83 | 8.57 | 4.10 | 8.20 | 4.17 |
| 23.53 | 4.19 | 25.19 | 4.30 | 26.11 | 3.90 | 23.73 | 4.13 | 25.74 | 4.16 |
| 23.55 | 4.17 | 50.50 | 4.30 | 48.64 | 3.81 | 23.86 | 4.14 | 25.77 | 4.15 |
| 48.99 | 4.11 | 69.82 | 4.09 | 71.72 | 3.78 | 48.04 | 3.98 | 48.72 | 4.04 |
| 71.61 | 4.08 | 69.88 | 4.09 | 71.76 | 3.75 | 71.57 | 4.05 | 72.45 | 3.92 |
| 71.68 | 4.04 | 117.75 | 4.10 | 119.12 | 3.65 | 71.60 | 4.03 | 72.49 | 4.05 |
| 118.75 | 3.88 | 221.18 | 3.97 | 240.67 | 3.58 | 118.17 | 3.84 | 119.08 | 3.95 |
| 240.37 | 3.80 | 221.25 | 4.00 | 240.70 | 3.60 | 239.15 | 3.78 | 238.95 | 3.60 |
| 240.38 | 3.77 | 477.58 | 3.84 | 475.68 | 3.26 | 239.20 | 3.75 | 239.00 | 3.60 |
| 477.72 | 3.32 | 477.62 | 3.85 | 475.73 | 3.20 | 507.45 | 3.65 | 473.22 | 3.18 |
| 477.75 | 3.40 | | | | | 507.48 | 3.61 | 473.25 | 3.20 |

**Table B.20**
**Winchester temperature studies (2 mg/L CLM residual at 24 hr)**

| EC 4°C | | EC 30°C | | $O_3$/BF 4°C | | $O_3$/BF 30°C | |
|---|---|---|---|---|---|---|---|
| Time (hr) | CLM residual (mg/L) | Time (hr) | CLM residual (mg/L) | Time (hr) | CLM residual (mg/L) | Time (hr) | CLM residual (mg/L) |
| Dose | 2.52 | Dose | 2.52 | Dose | 2.50 | Dose | 2.50 |
| 0.32 | 2.32 | 0.33 | 2.40 | 0.26 | 2.33 | 0.30 | 2.41 |
| 0.62 | 2.30 | 0.62 | 2.45 | 0.46 | 2.34 | 0.53 | 2.40 |
| 1.03 | 2.27 | 1.06 | 2.40 | 1.00 | 2.23 | 0.88 | 2.25 |
| 1.91 | 2.31 | 1.92 | 2.40 | 1.94 | 2.33 | 1.92 | 2.25 |
| 3.72 | 2.30 | 3.74 | 2.35 | 4.26 | 2.26 | 4.41 | 2.28 |
| 17.11 | 2.23 | 17.12 | 2.30 | 8.63 | 2.25 | 3.90 | 2.20 |
| 23.58 | 2.21 | 23.60 | 2.30 | 23.73 | 2.25 | 24.28 | 2.12 |
| 23.60 | 2.20 | 23.63 | 2.26 | 23.88 | 2.23 | 24.32 | 2.15 |
| 49.00 | 2.21 | 49.01 | 2.15 | 48.05 | 2.20 | 43.32 | 2.04 |
| 71.68 | 2.21 | 71.71 | 2.07 | 71.56 | 2.18 | 72.87 | 1.97 |
| 71.72 | 2.20 | 71.74 | 2.09 | 71.60 | 2.18 | 72.91 | 2.00 |
| 119.77 | 2.15 | 119.78 | 1.95 | 118.17 | 2.10 | 119.50 | 1.84 |
| 240.45 | 2.15 | 240.45 | 1.73 | 239.22 | 2.12 | 240.38 | 1.67 |
| 240.38 | 2.07 | 240.38 | 1.75 | 239.25 | 2.10 | 240.42 | 1.69 |
| 477.57 | 2.03 | 477.62 | 1.35 | 507.42 | 2.00 | 508.75 | 1.35 |
| 477.60 | 2.04 | 477.68 | 1.36 | 507.45 | 2.03 | 508.80 | 1.36 |

**Table B.21**
**Winchester EC pH study (2 mg/L CLM residual at 24 hr)**

| Time (hr) | CLM residual (mg/L) | Time (hr) | CLM residual (mg/L) |
|---|---|---|---|
| Dose | 2.51 | 53.93 | 2.23 |
| 0.45 | 2.36 | 76.35 | 2.00 |
| 0.93 | 2.40 | 76.38 | 2.03 |
| 2.07 | 2.40 | 123.42 | 1.98 |
| 4.10 | 2.36 | 245.30 | 1.75 |
| 24.75 | 2.25 | 245.33 | 1.75 |
| 24.83 | 2.22 | 480.35 | 1.57 |

**Table B.22**
**Winchester rechloramination study**

| EC BC | | Post-addition CLM | | HD CLM solution | |
|---|---|---|---|---|---|
| Time (hr) | CLM residual (mg/L) | Time (hr) | CLM residual (mg/L) | Time (hr) | CLM residual (mg/L) |
| Dose | 2.50 | Dose | 3.00 | 0.01 | **3.80** |
| 0.26 | 2.27 | 0.08 | 2.99 | 2.27 | 3.83 |
| 0.42 | 2.25 | 0.79 | 3.04 | 2.26 | 3.73 |
| 1.07 | 2.20 | 20.57 | 2.95 | 2.54 | 3.74 |
| 1.82 | 2.19 | 44.48 | 2.90 | 2.93 | 3.74 |
| 21.40 | 2.05 | | | 2.05 | 3.60 |
| 23.63 | 2.02 | | | 2.02 | 3.55 |
| 23.68 | 2.05 | | | 2.05 | 3.35 |
| 48.25 | 1.96 | | | 1.96 | 3.30 |
| 69.00 | 1.90 | | | 1.90 | 3.36 |
| 121.92 | 1.90 | | | 1.90 | 3.20 |
| 168.25 | 1.70 | | | | |
| 168.30 | 1.74 | | | | |

# APPENDIX C
# ANALYTICAL ERROR

Duplicate chlorine and chloramine residual measurements were taken for 23 percent of chlorine and 32 percent of chloramines measurements. The absolute average error and relative percent difference (RPD) were calculated according to equations C.1 and C.2 for both the chlorine and chloramine decay studies. A summary of the duplicate measurements and error analysis is shown in Table C.1 below.

$$\text{Absolute Error} = |M_1 - M_2| \tag{C.1}$$

$$\text{RPD} = \frac{|M_1 - M_2|}{M_{12}} \times 100\% \tag{C.2}$$

where $M_1$ = Chlorine/chloramine residual measurement
$M_2$ = Duplicate chlorine/chloramine residual measurement
$M_{12}$ = Mean value of $M_1$ and $M_2$

**Table C.1**
**Duplicate measurements and error analysis**

| | Duplicate sampling | | | Error analysis | |
|---|---|---|---|---|---|
| | No. of duplicate samples | No. of sample data points | Measurements duplicated (%) | Average absolute error (mg/L) | Average RPD (%) |
| $Cl_2$ | 178 | 758 | 23 | 0.042 | 3.9 |
| CLM | 236 | 735 | 32 | 0.018 | 0.68 |

# REFERENCES


APHA, AWWA, and WEF (American Public Health Association, American Water Works Association, and Water Environment Federation). 1998. *Standard Methods for the Examination of Water and Wastewater.* 20th ed. Washington, D.C.: APHA.

Bard, Y. 1974. *Nonlinear Parameter Estimation.* New York and London: Academic Press.

Biswas, P., C. Lu, and R.M. Clark. 1993. A Model for Chlorine Concentration Decay in Pipes. *Water Research*, 27(12):1715–1724.

Boccelli, D., M. Tryby, J. Uber, L. Rossman, M. Zierolf, and M. Polycarpou. 1998. Optimal Scheduling of Booster Disinfection in Water Distribution Systems. *ASCE Journal of Water Resources Planning and Management*, 124(2):99–111.

Boccelli, D.L., M.E. Tryby, J.G. Uber, and R.S. Summers. 2003. A Reactive Species Model for Chlorine Decay and THM Formation Under Rechlorination Conditions. *Water Research.*

Brezonik, P.L. 1994. *Chemical Kinetics and Process Dynamics in Aquatic Systems.* Boca Raton, Fla.: Lewis Publishers.

Clancy, J.L., Z. Bukhari, T.M. Hargy, J.R. Bolton, B.W. Dussert, and M.M. Marshall. 2000. Using UV to Inactivate Cryptosporidium. *Journal AWWA*, 92(9):97–104.

Clark, R.M. 1998. Chlorine Demand and TTHM Formation Kinetics: A Second-Order Model. *Journal of Environmental Engineering*, 124(1):16–24.

Clark, R.M., W.M. Grayman, J.A.Goodrich, R.A. Deininger, and K. Skov. 1994. Measuring and Modeling Chlorine Propagation in Water Distribution Systems. *ASCE Journal of Water Resources Planning and Management*, 120(6):871–887.

Clark, R.M., and M. Sivaganesan. 1998. Predicting Chlorine Residuals and Formation of TTHMs in Drinking Water. *Journal of Environmental Engineering*, 124(12):1203–1210.

Clark, R.M., and M. Sivaganesan. 2002. Predicting Chlorine Residuals in Drinking Water: Second-Order Model. *Journal of Water Resource Planning Manag*ement, 128(2):152–161.

Diehl, A.C., J.M. Symons, and S.W. Krasner. 1995. Factors Affecting Disinfection Byproduct Formation During Chloramination. *Proceedings, AWWA Annual Conference.* Denver, Colo.: AWWA.

Dotson, D., and G.R. Helz. 1985. Chlorine Decay Chemistry in Natural Waters. In *Water Chlorination.* Vol. 5. Edited by R.L. Jolley, R.J. Bull, W.P. Davis, S. Katz, M.H. Roberts Jr., and V.A. Jacobs. Chelsea, Mich.: Lewis Publishers.

Dugan, N.R., R.S. Summers, R.J. Miltner, and H.M. Shukairy. 1995. An Alternative Approach to Predicting Chlorine Residual Decay. *Proceedings, AWWA Water Quality Technology Conference.* Denver, Colo.: AWWA.

Duirk, S.E., B. Gombert, and R.L. Valentine. 2001. *Modeling Reactions of Monochloramine and Free Chlorine With NOM.* Iowa City, Iowa: University of Iowa.

Edzwald, J.K. 1994. Coagulation Concepts for Removal of Total Organic Carbon. *Proceedings, AWWA Water Quality Technology Conference.* Denver, Colo.: AWWA.

Fedorov, V.V. 1972. *Theory of Optimal Experiments.* New York and London: Academic Press.

Fleischacker, S.J., and S.J. Randtke. 1983. Formation of Organic Chlorine in Public Water Supplies. *Journal AWWA*, 75(3):132–138.

Gotoh, K. 1988. Residual Chlorine Concentration Decreasing Rate Coefficients for Various Pipe Materials. In *Proceedings, of 17th IWSA Water Supply Conference.* Vol. 7.

Haas, C.N., and S.B. Karra. 1984. Kinetics of Wastewater Chlorine Demand Exertion. *Journal WPCF*, 56(2):170–173.

Hao, O.J., A.P. Davis, and P.H. Chang. 1991. Kinetics of Manganese (II) Oxidation with Chlorine. *Journal of Environmental Engineering*, 117(3):359–374.

Harrington, G.W., Z.K. Chowdhury, and D.M. Owen. 1992. Developing a computer Model to Simulate DBP Formation During Water Treatment. *Journal AWWA,* 84(11)78–87.

Helz, G.R., D.A. Dotson, and A.C. Sigleo. 1983. Chlorine Demand: Studies Concerning Its Chemical Basis. In *Water Chlorination: Environmental Impact and Health Effects*, 4 (1):181–190. Edited by R.L. Jolley, W.A. Brungs, J.A. Cotruvo, R.B. Cumming, J.S. Mattice, and V.A. Jacobs. Ann Arbor, Mich.: Ann Arbor Science.

Hua, F., J.R. West, R.A. Barker, and C.F. Forster. 1999. Modeling of Chlorine Decay in Municipal Water Supplies. *Water Research*, 33(12):2735–2746.

Isabel, R.S., G. Solarik, M.T. Koechling, M.H. Anzek, and R.S. Summers. 2000. Modeling Chlorine Decay in Treated Waters. In *Proceedings, AWWA Annual Conference*. Denver, Colo.: AWWA.

Jadas-Hecart, A., A. El Morer, M. Stitou, P. Bouillot, and B. Legube. 1992. Modelisation de la Demande en Chlore dÕune Eau Traitee. *Water Research* 26(8):1073.

Jafvert, C.T., and R.L. Valentine. 1992. Reaction Scheme for the Chlorination of Ammoniacal Water. *Environment Science and Technology*, 26(3):577–586.

Jones, J.B. 2002. Mass Transfer Reactions and Decay Sinks for Disinfectants in Water Distribution Systems. Ph.D. dissertation, Old Dominion University, Norfolk, Va.

Karim, M.R., M. Abbaszadegan, and M.W. LeChevallier. 2003. Potential for Pathogen Intrusion During Pressure Transients. *Journal AWWA*, 95(5):134–146.

Kiene, L,. and Y. Levi. 1995. Influence of Corrosion on Chlorine Decay in Distribution Systems. *Proceedings, AWWA Water Quality Technology Conference*. Denver, Colo.: AWWA.

Kiene, L., W. Lu, and Y. Levi. 1993. Parameters Governing the Rate of Chlorine Decay Throughout Distribution System. In *Proceedings, AWWA 1993 Annual Conference*. Denver, Colo.: AWWA.

Knocke, W.R., H.L Shorney, and J.D. Bellamy. 1994. Examining the Reactions Between Soluble Iron, DOC, and Alternative Oxidants During Conventional Treatment. *Journal AWWA,* 86(1):117–127.

Koechling, M.T. 1998. Assessment and Modeling of Chlorine Reactions with Natural Organic Matter. Ph.D. dissertation, Department of Civil and Environmental Engineering, University of Cincinnati, Cincinnati, Ohio.

Langlais, B., D.A. Reckhow, and D.R. Brink (eds.). 1991. *Ozone in Water Treatment: Application and Engineering.* Cooperative Research Report, Awwa Research Foundation and Compagnie Generale des Eaux, Chelsea, Mich.: Lewis Publishers.

LeChevallier, M.W. 1999. The Case for Maintaining a Disinfectant Residual. *Journal AWWA*, 91(1):86–94.

Levenspiel, O. 1972. *Chemical Reaction Engineering*. 2nd ed. J. New York: Wiley and Sons.

Lu, W., L. Kiene, and Y. Levi. 1999. Chlorine Demand of Biofilms in Water Distribution Systems. *Water Research*, 33(3):827–835.

Maier, S.H., R.S. Powell, and C.A. Woodward. 2000. Calibration and Comparison of Chlorine Decay Models for a Test Water Distribution System. *Water Research*, 34(8):2301–2309.

Malley, J.P., J.P. Shaw, and J.R. Ropp. 1995. *Evaluation of By-products Produced by Treatment of Groundwaters With Ultraviolet Irradiation.* Denver, Colo.: AwwaRF.

Matheis, W., M.T. Koechling, R.S. Summers, and F. Frimmel. 1998. Influence of Drinking Water Treatment on Chlorine Decay Kinetics. *Proceedings, AWWA Annual Conference*. Denver, Colo.: AWWA.

Mathieu, L., J.C. Block, M. Dutang, J. Maillard, and D. Reasoner. 1993. Control of Biofilm Accumulation in Drinking-Water Distribution Systems. *Water Supply*, 11(3/4):365–376.

McClellan, J.N., D.A. Reckhow, J.E. Tobiason, J.K. Edzwald, and D.B. Smith. 2000. A Comprehensive Kinetic Model for Chlorine Decay and Chlorination By-Product Formation. In *Natural Organic Matter and Disinfection By-Products*. ACS Symposium Series 761. Edited by S.E. Barrett, S.W. Krasner, and G.L. Amy. Washington, D.C.: American Chemical Society.

Meckes, M.C., R. Haught, M. Dosani, R.M. Clark, and M. Sigvaganesan. 1999. Effect of pH Adjustments on Biofilm in a Simulated Distribution System. *Proceedings, AWWA Water Quality Technology Conference*. Denver, Colo.: AWWA.

Miltner, R.J., H.M. Shukairy, and R.S. Summers. 1992. Disinfection By-Product Control by Ozonation and Biotreatment. *Journal AWWA*, 84(11):53.

MWDSC (Metropolitan Water District of Southern California), James M. Montgomery Consulting Engineers, Inc., and Malcolm Pirnie, Inc. 1991: *Final Report: AWWA D/DBP Database and Model Project*. Contract No. 1-90.

Ozekin, K., R.L Valentine, and P.J. Vikesland. 1996. Modeling the Decomposition of Disinfecting Residuals of Chloramine. In *Water Disinfection and Natural Organic Matter: Characterization and Control*. Washington, D.C.: ACS Books.

Parent, A., S. Fass, M.L. Dincher, D. Reasoner, D. Gatel, and J.C. Block. 1996a. Control of Coliform Growth in Drinking Water Distribution Systems. *Journal CIWEM*, 10(December): 442–445.

Parent, A., S. Saby, M. Sardin, J.C. Block, and D. Gatel. 1996b. Contribution of biofilms to the chlorine demand of drinking water distribution systems. *Proceedings, AWWA Water Quality Technology Conference*. Denver, Colo.: AWWA.

Powell, J.C., N.B. Hallam, J.R.West, C.F. Forster, and J. Simms. 2000. Factors Which Control Bulk Chlorine Decay Rates. *Water Research*, 34(1):117–126.

Rajbhandari, A.N., M.T. Koechling, and R.S. Summers. 1999. Assessment of Chloramine Decay in Treated Waters; Impact of Treatment and Free Chlorine Contact Time. *Proceedings, AWWA Annual Conference*. Denver, Colo.: AWWA.

Rossman, L., R. Clark, and W. Grayman. 1994. Modeling Chlorine Residuals in Drinking-Water Distribution Systems. *Journal of Environmental Engineering*, 120(4):803–820.

Rossman, L.A. 1999. *EPANET2 Programmer's Toolkit* [Online]. Available: http://www.epa.gov/ORD/NRMRL/wswrd/epanet.html.

Rossman, L.A., R.A. Brown, P.C. Singer, and J.R. Nuckols. 2001. DBP Formation Kinetics in a Simulated Distribution System. *Water Research*, 35(14):3483–3489.

Shang, F., J. Uber, and M. Polycarpou. 2002. Particle Backtracking Algorithm for Water Distribution System Analysis. *Journal of Environmental Engineering*, 128(5):441–450.

Sharp, W.W., J. Pfeffer, and M. Morgan. 1991. In situ chlorine decay rate testing. *Proceedings, Water Quality Modeling in Distribution Systems Conf*erence. Denver, Colo.: AwwaRF.

Shorney, H.L., and W.R. Knocke. 1996. Bench Scale Evaluation of Manganese Stability in the Presence of Chlorine and Combined Chlorine. *Proceedings, AWWA Water Quality Technology* Conference. Denver, Colo.: AWWA.

Shukairy, H.M. 1994. The Control of Disinfection By-Product Formation by Ozonation and Biotreatment. Ph.D. dissertation. University of Cincinnati, Cincinnati, Ohio.

Shukairy, H.M., R.J. Miltner, and R.S. Summers. 1994. Bromide's Effect on DBP Formation, Speciation, and Control: Part 1, Ozonation. *Journal AWWA,* 86(6):72–87.

Shukairy, H.M., R.J. Miltner, and R.S. Summers. 1995. Bromide's Effect on DBP Formation, Speciation, and Control: Part 2, Biotreatment. *Journal AWWA,* 87(10):71–82.

Summers, R.S. 1986. *Activated Carbon Adsorption of Humic Substances: Effect of Molecular Size and Heterodispersity.* Ph.D. dissertation. Department of Civil and Environmental Engineering, Stanford University, Palo Alto, Calif.

Summers, R.S., S.M. Hooper, H.M. Shukairy, G. Solarik, and D. Owen. 1996. Assessing DBP Yield: Uniform Formation Conditions. *Journal AWWA*, 88(6):80–93.

Summers, R.S., G. Solarik, A. Rajbhandari, and M. Koechling. 2001. Modeling Chloramine Residuals in Treated Waters. *Proceedings, AWWA Annual Conference*. Denver, Colo.: AWWA.

Van Hoof, F. 1992. Identifying and Characterizing Effects of Disinfection Byproducts. In *Disinfection Alternatives for Safe Drinking Water.* Edited by E.A. Bryant, G.P. Fulton, and G.C. Budd. New York: Van Nostrand Reinhold.

Vasconcelos, J.J., P.F. Boulos, L.A. Rossman, W.M. Grayman, R.M. Clark, and J.A. Goodrich. 1995. Characterizing and Modeling Chlorine Decay in Distribution Systems—A Summary. In *Proceedings, AWWA Annual Conference.* Denver, Colo.: AWWA.

Vasconcelos, J.J. W.M. Grayman, L. Kiene, O. Wable, P. Biswas, A. Bhari, L.A. Rossman, R.M. Clark, and J.A. Goodrich. 1996. *Characterization and Modeling of Chlorine Decay in Distribution Systems.* Denver, CO.: AwwaRF.

Vasconcelos, J.J., L.A. Rossman, W.M. Grayman, P.F. Boulos, and R.M. Clark. 1997. Kinetics of Chlorine Decay. *Journal AWWA*, 89(7):54–65.

Vikesland, P.J., K. Ozekin, and R.L Valentine. 1998. Effect of Natural Organic Matter on Monochloramine Decomposition: Pathway Elucidation Through the Use of Mass and Redox Balances. *Environment Science and Technology*, 32(10):1409–1416.

Vikesland, P.J., K. Ozekin, and R.L. Valentine. 2001. Monochloramine Decay in Model and Distribution System Waters. *Water Res*earch, 35(7):1766–1776.

Vikesland, P.J., R.L. Valentine, and K. Ozekin. 1996. Application of Product Studies in the Elucidation of Chloramine Reaction Pathways. In *Water Disinfection and Natural Organic Matter: Characterization and Control.* Washington, D.C.: American Chemical Society.

Wable, O., N. Dumoutier, J.P. Duguet, P.A. Jarrige, G. Gelas, and J.F. Depierre. 1991. Modeling Chlorine Concentrations in a Network and Applications to Paris Distribution Network In *Proceedings, Water Quality Modeling in Distribution Systems Conf*erence. Denver, Colo.: AwwaRF.

White, G.C. 1992. *Handbook of Chlorination and Alternative Disinfectants*. 3rd ed. New York: Van Nostrand Reinhold.

Wilczak, A., L.L Hoover, and H.H. Lai. 2003. Effects of Treatment Changes on Chloramine Demand and Decay. *Journal AWWA*, 95(7):94–106.

Wolfe, R.L., N.R. Ward, and B.H. Olson. 1984. Inorganic Chloramines as Drinking Water Disinfectants: A Review. *Journal AWWA*, 76(5):74–88.

Zierolf, M., M. Polycarpou, and J. Uber. 1998. Development and Auto-Calibration of an Input-Output Model of Chlorine Transport in Drinking Water Distribution Systems. *IEEE Transactions on Conrol Systems Technology*, 6(4):543–553.

# ABBREVIATIONS

| | |
|---|---|
| ACS | American Chemical Society |
| ANOVA | analysis of variance |
| AWWA | American Water Works Association |
| AwwaRF | Awwa Research Foundation |
| BDOC | biodegradable dissolved organic carbon |
| BF | biofiltration |
| C | degrees centigrade |
| CAP | Central Arizona Project |
| $CD_t$ | chlorine demand at time t |
| $CLMD_t$ | chloramine demand at time t |
| cm | centimeter |
| CT | residual disinfectant concentration (mg/L) x time (min) |
| DBP | disinfection byproduct |
| D/DBP | disinfection and disinfection byproduct |
| DI | deionized |
| DLL | dynamic link library |
| DOC | dissolved organic carbon |
| DPD | N,N-diethyl-p-phenylenediamine |
| DWDS | drinking water distribution system |
| EC | enhanced coagulation |
| ft | feet |
| g | gram |
| GAC | granular activated carbon |
| gpm | gallons per minute |
| hr | hours |
| I/O | input/output |
| IID | independent and identical distribution |
| km | kilometer |
| L | liter |
| ln | natural log (base e logarithm) |
| M | molar (mol/L) |
| M/DBP | microbial and dsifection byproduct |
| mg | milligrams |
| mgd | million gallons per day |
| mg/L | milligrams per liter |

| | |
|---|---|
| min | minutes |
| mJ | millijoules |
| mL | milliliters |
| mm | millimeters |
| mW | milliwatts |
| μS | microsiemens |
| | |
| NF | nanofiltration |
| nm | nanometers |
| NOM | natural organic matter |
| NPA | Network Path Analysis |
| NPNV | Newport News Virginia |
| NTU | Nephelometric Turbidity Units |
| | |
| $O_3$ | ozone |
| ODE | ordinary differential equation |
| | |
| PBA | particle backtracking algorithm |
| | |
| $r^2$, $R^2$ | correlation coefficient |
| RMSE | root mean square error |
| RO | reverse osmosis |
| rpm | revolutions per minute |
| | |
| SCADA | Supervisory Control and Data Acquisition |
| sec | seconds |
| SM | Standard Method |
| SUVA | specific ultraviolet absorbance to dissolved organic carbon ratio |
| | |
| T&E | USEPA Test and Evaluation Facility |
| TCR | Total Coliform Rule |
| TDS | total dissolved solids |
| THM | trihalomethane |
| TOC | total organic carbon |
| TSUVA | specific ultraviolet absorbance to total organic carbon ratio |
| | |
| USEPA | United States Environmental Protection Agency |
| UV | ultraviolet |
| UVA | ultraviolet absorbance |
| | |
| WTP | water treatment plant |

IWA Publishing's authorised EU representative for General Product Safety Regulations is Diane D'Arras, 15 rue Duret, 75116 Paris, France, e-mail: safety@iwap.co.uk.

Printed and bound by CPI Group (UK) Ltd, Croydon, CR0 4YY
17/09/2026
02218799-0002